MATERIALS SCIENCE AND TECHNOLOGIES

ELECTRO-OPTIC EFFECTS IN THERMOTROPIC LIQUID CRYSTALS

MATERIALS SCIENCE AND TECHNOLOGIES

Additional books in this series can be found on Nova's website
under the Series tab.

Additional e-books in this series can be found on Nova's website
under the e-book tab.

MATERIALS SCIENCE AND TECHNOLOGIES

ELECTRO-OPTIC EFFECTS IN THERMOTROPIC LIQUID CRYSTALS

HRISTO PETKOV HINOV

nova publishers
New York

Library of Congress Cataloging-in-Publication Data

ISBN: 978-1-63117-507-7

Published by Nova Science Publishers, Inc. † New York

This book is dedicated to Rayna, Nadezhda and Elitsa

CONTENTS

PREFACE

The material included in this work is intended for scientists working both in the science and the application of the liquid crystals. We selected some of our more interesting and useful results, which, presumably can be elaborated further. We understand that this material is a pea in the ocean of liquid crystal knowledge, but we believe that the results, presented here, will be of interest to many researchers from the liquid crystal community.

Hristo Petkov Hinov, PhD
Associate Professor
Institute of Solid State Physics, Bulgarian Academy of Sciences
Laboratory of Liquid Crystals
P. O. Box 33, Sofia 1784, Bulgaria
hinov@issp.bas.bg

ABSTRACT

New methods and phenomena, including thermo-effects, are suggested for a deeper understanding of some interactions between the bulk and the surface of the smectics. The smectic-A or smectic-C phases grow from a nematic or unwinded large-pitch cholesteric oriented in the middle part of the cells and from strongly-deformed nematic or unwinded large-pitch cholesteric layer in the boundary regions of the liquid crystal cells. Such a way of growing of the smectics has two advantages: first, it shows the texture fitting between the surface and bulk alignment of the smectics, sometimes unknown, and permits in some cases the approximate determination of the surface energy of smectics and second, it is useful for application in the laser-addressed liquid crystal devices. On the other hand, the experimental results obtained with the smectic-C permit the evaluation of certain material parameters. The present work includes also the electro-optic behavior of the large-pitch cholesteric at a new, strong-weak or weak-weak anchoring as well as the hysteretic behavior leading to the appearance or disappearance of various large-pitch textures. One of the most interesting electro-optic effects, discovered in the large-pitch cholesterics, is the observation of polar unidirectional spirals under a DC voltage application. Finally, we draw attention to the discovery of a new chiral liquid crystal with unusual electro-optic behavior composed from a mixture of a nematic and a ferroelectric smectic-C* liquid crystals. We have experimentally studied the electro-optic dynamic hysteretic behavior of weakly-anchored twisted nematic layer as well. The flexoelectric deformations were identified except by additional application of a high-frequency orienting electric field and, qualitatively, by the simultaneous application of DC and low-frequency (several hertz) electric fields. In this way, for the first time, unusual amplification was observed of the low-frequency flexoelectric oscillations. The „gradient flexoeffect" can be realized not only by the accumulation of ions in the nematic layer, but also by inhomogeneity of the electric field, created by the geometry of the liquid crystal cell. For example, we regarded the flexoeffect due to the wedge-shape of the liquid crystal layer. Finally, we focus our attention on the polar flexoeffect, without taking into account the second-order elasticity.

The pulsations of the light transmitted through a nematic layer with a homeotropic anchoring and a free-slipping boundary are again considered, and new ideas are developed. The material parameters are of a great importance for the theory and application of the liquid crystals. Due to this reason, we have suggested new methods for their measurement, some cases being illustrated by examples. Finally, we proposed the Becke's method for measurement of the optical index of various solids to be combined with the use of a monodomain liquid crystal in its nematic or isotropic phase.

INTRODUCTION: BRIEF HISTORICAL NOTES ON THE DISCOVERY AND STUDY OF THE LIQUID CRYSTALS

The liquid crystals are discovered by the Austrian biologist Friedrich Reinitzer in 1888 with a significant assistance of the German physicist Otto Lehman who has been universally recognized authority in that time in the field of microscope technique [1-5]. The original work of Reinitzer written in German, has been translated in English and re-published in the journal "Liquid crystals" due to the extraordinary importance of the paper and on the occasion of the 100th anniversary of the discovery of the liquid crystals [6, 7]. By irony of fate, Reinitzer not only discovered the new, fourth state of matter as is the studied by him cholesterol benzoate, but also observed for the first time the extraordinary interesting complex blue phase of liquid crystals [8, 9] which was recognized many years later [7]. The term "liquid crystal" has been suggested for the first time by Lehman. Later, in the beginning of 20th century, the famous scientist George Friedel proposed two, more specific, terms, related to the different states of the liquid crystal, namely, "nematic" and "smectic". Before 1941 the liquid crystals have been intensively studied in Germany (the University of Halle), in France (The University of Paris) and in Russia (The University of Kharkov). The first electro-optic static effect and the first electro-hydrodynamic effect (electro-hydrodynamic scattering of light) have been studied first in Russia and in Germany. The early theoretical works of Oseen, Frank and Ericksen [10] on the static theory of liquid crystals and of Ericksen-Leslie for the hydrodynamic theory of liquid crystals [10,11], exercised enormous influence on the study, in general, of the liquid crystals [12-15]. A further stimulus for the study of liquid crystals EXERTEDthe publications of the Americal scientist George Heilmeier [16] who in a number of papers proved that the effect "dynamic scattering of light" by nematics can be applied in the liquid crystal displays, and even in the television. Now the interest in the liquid crystals has moved to the electric-field effects since they possess better electro-optic characteristics, their times of reaction are much shorter and they live much longer(about 100 thousands hours). At present there are discovered and applied more than100 effects in the liquid crystal displays (see, for instance [17-27]). We should also point out two papers about the determination of heroin by means of the pitch of induced cholesteric mesophases [28] and a new form (now widely used) of the liquid crystal touch switch [29]. As a consequence of the effort of chemists, physicists and engineers, more than hundred thousand papers both theoretical and experimental have been published. More than hundreds of books have been published, more in the research, less in the practice. We should note two wonderful books, published by de Gennes [31] and by Chandrasekhar [32]. In addition, more than several hundred important reviews have been published, among them [33-35]. Furthermore, over hundreds of collections of important papers have been published, among them [36]. We will finish with part of the" Recollections of an old student" [36] by de Gennes, Nobel Prize winner in Physics for 1991 [37]: "The story of liquid crystals is an Odyssey…"

REFERENCES

[1] Kelker, H. *Mol. Cryst. Liq. Cryst.* 1973, 21, 1-48.

[2] Kelker, H. In *Zehn Arbeiten über Flüssigkristalle; Wissenschaftliche beitrage* 1986/52, N 171, Halle (Saale), Martin-Luther Universitat 1986; pp 193-215.

[3] Kelker, H. *Mol. Cryst. Liq. Cryst.* 1988, 165, 1-43.

[4] Reinitzer, F. *Monatshefte für Chemie* 1888, 9, 421-441; English Translation in "Liquid crystals" 1988, 5, 7-18.

[5] Lehman, O. *Z. Phys. Chem.* 1889, 4, 462-472.

[6] Stegemeyer, H. *Angew. Chem.* 1988, 100, 1640-1642.

[7] Stegemeyer, H. *Liq. Cryst.(UK)* 1989, 5, 5-6.

[8] Stegemeier, H.; Blümel, Th.; Hiltrop. K.; Onusseit, H.; Porsch. F. *Liq. Cryst.(UK)* 1986, 1, 3-28.

[9] Seideman, T. *Rep. Prog. Phys.* 1990, 53, 659-705.

[10] Ericksen, J. L. In *Advances in Liquid crystals*; Brown, G.; Ed.; Academic Press, Inc. New York- San Francisco-London, 1976; vol. 2; pp 233-298.

[11] Leslie, F. M. In *Advances in Liquid crystals*; Brown, G.; Ed.; Academic Press, Inc. New York, 1979; vol. 4; pp 1-81.

[12] Carlsson, T.; Leslie, F. M. *Liq. Cryst.(UK)* 1999, 26, 1267-1280.

[13] Frank, F. C. *Phil. Trans. R. Soc. Lond. A* 1983, 309, 71-75.

[14] Nakagawa, M. *Liq. Cryst.(UK)* 1997, 23, 561-567.

[15] Derfel, G. *Liq. Cryst.(UK)* 1998, 24, 829-834.

[16] Heilmeier, G. H.; Zanoni, L. A.; Barton, L. A. *Proc. IEEE* 1968, 56, 1162-1171.

[17] Goodman, L. A. *RCA Rev.* 1974, 35, 447-461.

[18] Berreman, D. W. *Phil. Trans. R. Soc. Lond.A* 1983, 309, 203-216.

[19] Bahadur, B. *Mol. Cryst. Liq. Cryst.* 1983, 99, 345-374.

[20] Haas, W. E. *Mol. Cryst. Liq. Cryst.* 1983, 94, 1-31.

[21] Castellano, J. A. *Mol. Cryst. Liq. Cryst.* 1988, 165, 389-403.

[22] Schadt, M. *Liq. Cryst.(UK)* 1989, 5, 57-71.

[23] Adachi, K. *Sol. St. Technol.* 1993, January, 63-71.

[24] Schadt, M. *Ann. Rev. Mat. Sci.* 1997, 27, 305-379.

[25] Abbate, G.; Otòn, J. M. *Adv. Mater.* 2000. 12, 459-467.

[26] Crossland, W. A.; Wilkinson, T. D.; Manolis, I. G.; Redmond, M. M.; Davey, A. B. *Mol. Cryst. Liq. Cryst.* 2002, 375, 1-13.

[27] Crossland, W. A.; Clapp, T. V.; Wilkinson, T. D.; Manolis, I. G.; Georgiou, A. G.; Robertson. B. *Mol. Cryst. Liq. Cryst.* 2004, 413, 2499-2519.

[28] Bertocchi, G.; Gottarelli, G.; Prati, R. *Talanta* 1984, 31, 138-140.

[29] Bradshaw, M. J.; Hilsum. C.; Raynes, E. P. *El. Lett.* 1982, 18, 1106-2208.

[30] Ondris-Crawford, R. J.; Crowford, G. P. ; Doane, J. W. *Am. J. Phys.* 1995, 63, 781-788.

[31] De Gennes, P. G. *The Physics of Liquid crystals* ; Clarendon Press: Oxford, UK, 1974.
[32] Chandrasekhar, S. *Liquid crystals*; Cambridge University Press, UK, 1977.
[33] Stephen, M. J.; Straley, J. P. *Rev. Mod. Phys.* 1974, 46, 618-704.
[34] Saupe, A. *Ann. Rev. Phys. Chem.* 1973, 24, 441-471.
[35] Small. D. M. *J. Coll. Int. Sci.* 1977, 58, 581-602.
[36] *Dynamics and Defects in Liquid crystals (A Festschrift in Honor of Alfred Saupe)*; Cladis, P. E.; Palffy-Muhoray, P.; Eds.; Gordon and Breach Sci. Publ., OPA, 1998.
[37] De Gennes, P. G. *Soft Matter* (This lecture was delivered 9 December 1991, on the occasion of the presentation of the 1991 Nobel Prize in Physics) *Rev. Mod. Phys.* 1992, 64, 645-648; *Uspechi Fiz. Nauk (*Russia*), In Russian* 1992, 162, 125-132.

Chapter 1

FORMATION OF THE SMECTIC-A LIQUID CRYSTAL FROM STRONGLY-DEFORMED IN THE BOUNDARY REGIONS NEMATIC

1.1. GENERAL QUESTIONS CONSERNING THE STRUCTURE OF THE SMECTIC-A AND THE NEMATIC-SMECTIC-A PHASE TRANSITION

Since the discovery of the first liquid crystal in 1888 by the Austrian biologist Friedrich Reinitzer with the considerable assistance of the German physicist Otto Lehman, who in that time had been universally recognized authority in the field of microscope technique [1–3], thousands and thousands of liquid crystals have been synthesized among them nematics, short-pitch cholesterics and various kinds of smectics such as smectic-A, smectic-C, etc. [4–7].

The smectic-A liquid crystal, as well as the other smectics, are substances with a structure and properties peculiar both to liquids and solids [8–11]. The molecules of the smectic-A phase are arranged in layers with a thickness, approximately one half length of the molecule, which is a typical property of the simpler smectics-A (certainly this is not true for the more complex in their structure smectics-A: now a number of smectics-A are discovered and classified by de Vries [12]). A second important characteristic feature is that the centers of gravity of the molecules do not possess an order inside the smectic-A layers, i.e., such a long-range layer represents a two-dimensional liquid. A third basic characteristic feature, which according to de Gennes [13] follows from the first two, is the unidirectional alignment of the molecules. It is clear that the smectic-A is unidirectional medium with an optical axis which is perpendicular to the planes of the smectic-A layers (this axis usually coincides with the axis Z of the chosen coordinate system). The assumption of a high symmetry around the axis of rotation OZ does not contradict to all hitherto known experimental data. Furthermore, the directions Z and –Z are equivalent which shows that the simple smectic-A is unipolar (now many polar smectics-A are known, but we do not focus our attention on them). In summary, the smectic-A, in contrast to the nematic, has exceptionally orientational order and translational one, i.e., it combines the properties of a two-dimensional liquid and of one-dimensional crystal. (In the following, the description of the structure and properties of the smectic-A and its behavior around the important nematic- smectic-A phase transition is based on the book of de Gennes [13] and reviews, written by leading scientists in this field of research [14-24]).

The microscopic theory of the smectic-A has been developed for the first time by Kobayashi [25] and, independently, by McMillan [26]. According to Chandrasekhar [27], McMillan has suggested a simple and elegant description of the smectic-A phase modifying the Maier-Saupe [28] theory for nematics with additional translational periodicity in the layer structure. The smectic-A layers can be phenomenologically described with the aid of complex order parameter designated by the complex function $\Psi(z)$. The amplitude of Ψ describes the straight of the smectic-A layers and the phase of Ψ describes the position of the layers and the distance between them. The phase of Ψ is a constant on the surface of the smectic-A layer and it is larger by 2π on the next smectic-A layer. On the other hand, the strength of the interaction between the thermal fluctuations of the nematic director n and the smectic-A order, represented by the complex function Ψ, determines definitely the order of the nematic-smectic-A transition. In other words, the wider is the nematic interval, the more probable is the ratio T_{NA}/T_{NI} to be smaller than 0.87 (McMillan) and the more probable is the nematic-smectic-A phase transition to be of second order. The smectic-A order parameter is important not only near the nematic-smectic-A phase transition which would be the ideal case, but also far away from the transition due to the interactions of both thermal fluctuations. The order parameter of the nematic phase cannot reach saturation when the nematic interval is too narrow. At these circumstances, the growing of the smectic-A order parameter leads also to the growing of the nematic order parameter, i.e., when the connection between the two order parameters is strong enough, the nematic-smectic-A phase transition is of a first order. If the nematic phase exists in a wider temperature interval, the interaction between the two order parameters considerably decreases and the nematic-smectic-A phase transition becomes of a second order [29].

The best description of the nematic-smectic-A phase transition can be found, for instance, in the pioneering work of de Gennes [30] (see also the References from 31 to 33). He wrote the density of the free energy of the nematic phase near the nematic-smectic-A phase transition in the form of Landau-Ginzburg:

$$F = F_0 + \alpha|\psi|^2 + (\beta/2)|\psi|^4 + (1/2M_v)\,|\partial_z\psi|^2 + (1/2)M_t|(\partial_x + iq_0 n_x)\psi|^2$$
$$(1/2)M_t|(\partial y + iq_0 n_y)\psi|^2 + (K_1/2)[\partial_x n_x + \partial_y n_y]^2 + (K_2/2)[\partial_x n_y - \partial_y n_x]^2$$
$$+ (K_3/2)[(\partial_z n_x)^2 + (\partial_z n_y)^2] \tag{1.1}$$

where for the density of the smectic-A layers one accepts the expression

$$\rho = \rho_0\left[1 + \mathrm{Re}(\psi e^{iq_0 z})\right] \tag{1.2}$$

(see Figure 1.1):

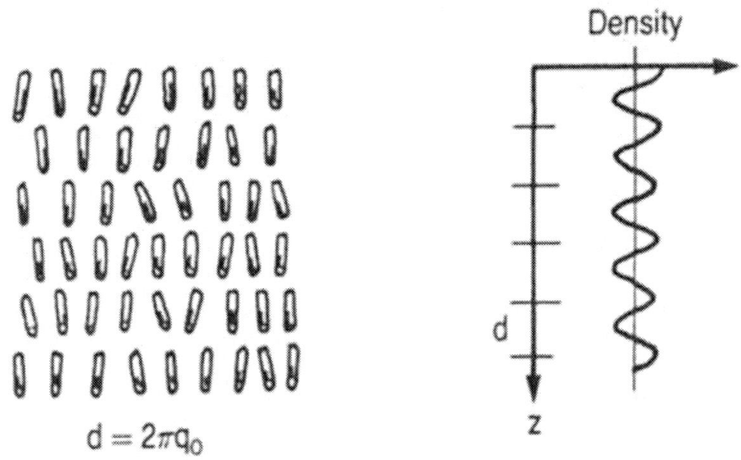

Figure 1.1. Schematic representation of the structure of smectic-A phase.

where $d=2\pi/q_0$ is the distance between the smectic-A layers and $k=|\Psi|\exp(iq_0u)z$ is a complex two-component function describing the smectic-A order of alignment. The phase of k, expressed by q_0u, represents the deformations of the smectic-A layers in OZ direction, parallel to the undisturbed director of the nematic. The energy in form (1.1) has been written by de Gennes on the basis of the symmetric properties of Ψ. Several very important conclusions follow from the kind of the elastic energy, written by de Gennes in the form (1.1):

(a) The elastic constants of twist K_{22} and bend K_{33} increase very sharply on approaching the nematic-smectic-A phase transition according to the following relations:

$$K_2 = K_2^0 + (kT/24\pi)\, q_0^2 \left(\varsigma_\perp^2 / \varsigma_{II}\right)$$
$$K_3 = K_3^0 + (kT/24\pi)\, q_0^2\, \varsigma_{II} \qquad (1.3)$$

where K_i^0 are the undisturbed elastic constants of the nematic, k is the constant of Boltzmann, q_0 is the density vector of the smectic-A phase. $\zeta_{\|} = (2\alpha/M_v)^{-1/2}$ and $\zeta_\perp = (2\alpha/M_t)^{-1/2}$ are the longitudinal and transversal correlation length of the smectic-A short range order, M_v and M_t are the components of the "mass tensor" along and perpendicular to undisturbed director, respectively.

(b) The nematic-smectic-A phase transition temperature decreases in a strongly-deformed nematic.

The splay deformation does not influence the value and kind of the nematic-smectic-A phase transition. The pre-transitional critical phenomena in the nematic phase on approaching the smectic-A phase lead to the creation of smectic-A groups, or "clusters", with correlation lengths $\zeta_{\|}$ (parallel to the director) and $\zeta_\perp$ (perpendicular to the director). These „clusters" can be represented roughly as "smectic islands" embedded into "nematic sea".

More precisely, however, they can be represented as small nematic regions fluctuating "toward" and "away" from a smectic configuration, e.g., fluctuations of the smectic-A order of arrangement [34-37]. In addition to this, one notes a considerable increase in some of the elastic constants and viscosities of the nematic. For instance, the increase in the magnitude of the elastic constants of twist and bend near the nematic-smectic-A phase transition is different and the ratio $\delta K_{33}/\delta K_{22}$ can reach from 8 to several orders of magnitudes [38]. As noted, the cybotactic groups influence also the dynamic behavior of the liquid crystal near the nematic-smectic-A phase transition and in particular lead to the critical increase of some viscous coefficients of the nematic, included the rotational viscosity coefficient [39, 40]. Now there is a commonly accepted opinion that the smectic cluster appears in the nematic phase as a rotational ellipsoid with a long axis, perpendicular to the smectic-A layers. On approaching the phase transitional temperature T_c, the length of this axis determining the number of smectic layers, is critically increased, whereas the lateral growth of the length of the layers is slower. In two fundamental works Halperin and Lubensky [31] and Halperin, Lubensky and Ma [32] theoretically predicted that the fluctuations of the director in the nematic phase as well as their interactions with the smectic-A order always lead to a nematic – smectic-A phase transition of a first order, which at least should be a weak first order. These two basic papers were followed by a number of theoretical and experimental studies [41-47]. According to Rosciszewski [41], most of the experimental facts point to a mixed behavior with features both of the mean-field model (McMillan) and of the superfluid helium model (de Gennes). The key to the solution of this very interesting and complex phase transition can probably be found in the explanation of Lubensky and Chen [48] that the type of the phase transition changes with the temperature, i.e., near the nematic – smectic-A phase transition it is of a second order and very near to the same transition (for instance, $< 3\ mK$), it is of first order [49, 50]. We stress here that our experimental results are explained with the aid of the model proposed by de Gennes [30]. The decrease of the phase transitional nematic – smectic-A temperature was proved in our experiment with the growing of the smectic-A from a strongly deformed nematic with deformations of splay-bend. Furthermore, our experiments clearly showed that this growing occurs when the phase transition is of a second order or a slightly first-order. This experimental fact was also confirmed by Madhusudana and Srikanta [51]. Cladis and Torza [52] were the first to show experimentally that the phase transitional nematic – smectic-A temperature decreases when the smectic-A grows from a strongly deformed nematic layer. On the other hand, Lubensky [53] has shown theoretically that the temperature of the cholesteric – smectic-A phase transition decreases with a magnitude, which is in the range of $10^{-3}\ T_{NA}$. This small temperature difference is due to the twist, existing in the cholesteric phase. This theoretical result has been experimentally confirmed 20 years later by Crandal et al., [54]. In addition, Madhusudana and Srikanta showed experimentally that the growth of the smectic-A phase from a deformed, for instance twisted nematic with a nematic – smectic-A phase transition of second order, or a slightly first order, is accompanied by the simultaneous existence of both phases: the nematic and smectic-A and to a decrease of the phase transitional temperature by several degrees.. According to the theory of de Gennes [30] this result shows that the following inequality is performed:

$$K < 1/\sqrt{2}$$

where

$$K = \lambda / \xi$$

λ is the penetration depth of |rot**n**| and ξ is the coherence length showing the influence of the perturbations on the smectic order. If the above inequality changes to the opposite:

$$K > 1/\sqrt{2}$$

both phases, nematic and smectic-A, are separated by the phase of Shubnikov [30, 51]. According to de Gennes, materials possessing a nematic – smectic-A phase transition of second order or nearly second order (i.e., a very weak first order) are materials of a first kind and the parameter of Ginzburg-Landau satisfies the first inequality. This question has also been considered by Ranganath [55] when disclinations and dislocations exist in the smectic-A or smectic-C phases, and by Durand [56] who has considered antagonistic nematic and smectic-A surface anchoring. Tamblyn et al., [57] have also considered the melting of the smectic-A phase at strong deformations. The simultaneous existence of the nematic and smectic-A phases in the form of stripe domains (domains of Cladis-Torza) have been observed by many authors and there is indication for the critical increase of the elastic constants of twist and bend. These stripes have not been observed when the smectic-A phase grows from a purely twisted nematic [51]. Chu and McMillan [58] and Majoros et al., [59] and Gooder et al., [60] on the one hand and Allender et al., [61] on the other hand made an attempt to explain alternatively the stripe domains by considering one-dimensional Fréedericksz transition of a first order [58] or a two-dimensional Fréedericksz transition [59, 61]. In addition, Chevallard and Clerk [62] claimed in 2001 that the problem with the Cladis-Torza's stripes can be resolved with a non-linear elastic theory (see also References [63] and [64]). Stripe domains in another geometry have also been observed by Pergamenshchik et al., in 2008 [65].

All the liquid crystals used in our experiments in the period 1981-1990, such as CBOOA (cyanobenzylidene-octylaniline) [66,67], NPOB (4-nitrophenyl-4'-octyloxybenzoat) [68], 8CB (4-n-octyl-4'-biphenyl) [69, 70] show critical behavior of the various material parameters and a very small enthalpy of the nematic – smectic-A transition, usually smaller than kT (see, for instance References [71] and [72]).. The basic conclusion is that the elastic constants of bend K_{33} and twist K_{22} increase near the nematic – smectic-A phase transition in such a way that the increase of K_{33} is bigger relative to that of K_{22}. As mentioned above, an important question at discussion of our results obtained, is the value of the latent heat accompanying the nematic – smectic-A phase transition which for the liquid crystal CBOOA is, for instance, 0.084 kJ/mole [73](see also Reference [74]). Bata et al., [75] measured the latent heat accompanying the nematic-smectic-A phase transition for the liquid crystal NPOB and have obtained 0,092 kJ/mol. This phase transition according to their conclusions has been of a slightly first order. Coles and Strazielle [76] measured a latent heat of 0.040 kJ/mole for the liquid crystal 8CB and pointing out a nematic – smectic-A phase transition of a very weak first order. Kasting et al., [77], and Navard and Cox [78] measured a very weak latent heat varying between 0 and 0.010 kJ/mole and pointing to a second order nematic – smectic-A phase transition. Later, our experimental results also demonstrated that the liquid crystal 8CB has a nearly nematic - smectic-A phase transition of a second order [69, 70]. Contrary to the claims of some authors, our experimental results clearly show that the matching of the strongly deformed nematic boundary regions with the well-oriented bulk smectic-A is realized by the creation of a system of confocal domains, typical for liquid crystals such as 8CB, which rather shows nematic – smectic-A phase transition of a second order.

For completeness, we point out some papers dealing with the pretransitional behavior of the various material parameters at the nematic-smectic-A phase transition for the liquid crystal CBOOA [79–98], NPOB [99–101], 8CB [102–107] and CBOOA plus 8CB [108,109]. It is evident that due to de Gennes, Nobel prizer for 1991 [84], we know now much about this very interesting phase transition.

1.2. Experimental Results

Most of the experimental results concerning the smectic-A were obtained with the aid of plane-parallel cells consisting of two glass plates which were covered from the one side by a very thin transparent conductive layer, two isolation sheets separating the glass plates and determining the thickness of the liquid (it varied between 5 μm and 200 μm in the various experiments) and two metallic side holders (see Figure 1.2). The cell, shown in Figure 1.2 was prepared in 1979 in the University of Paris and its stability is ensured by two lateral massive metal holders and four screws. The preparation of such a cell has several advantages, namely the possibility of achieving high temperature gradients in the electrode planes, easy study of a large area (several cm^2) of the liquid crystal and easy photography of the liquid crystal structures with objectives having small focal length and a magnification above 100.

Figure 1.2. A view of the liquid crystal cell prepared.

The voltage was applied to the conducting sides of the glass plates (a normal electric field) for the experiments performed with the liquid crystals CBOOA and 8CB which were introduced in the cells by capillary forces (the so-called capillary filling of the liquid crystal [110]). For the case of the liquid crystal NPOB, the isolating sheets were replaced by conductive metal sheets or by usual wires and the glass plates were not conducting (an electric field in the plane of the glasses). The initial alignment of the nematic was either planar or homeotropic. The planar orientation was achieved by standard methods of alignment such as evaporation of a very thin layer of SiO [111-116] or a rubbed thin polymeric layer, for instance polyvinyl alcohol (PVA) [117-122] (the process of rubbing will be considered in detail in chapter 4). The homeotropic orientation of the nematic or smectic-A director was achieved either by cleaning very carefully the NESA coated glass plates Electrodes [123, 124] or by using Surfactants [125] such as Lecithin [126-128]. The change of the temperature [129-133] was achieved by two methods. The first one was simpler and was achieved by heating the whole cell up to the isotropic phase of the liquid crystal followed by a free cooling of the sample. The temperature change was facilitated by the side massive metal holders absorbing a big amount of the thermal energy which they gradually transferred to the liquid crystal. The second method was performed by using a thermo-regulator and a thermostatic plate with a wide range of regulation of the temperature with a velocity between 0.1^0 C/min, 1^0 C/min, 3^0 C/min, and 10^0 C/min. The liquid crystal cells prepared by the above-mentioned procedure and possessing different alignment: planar, homeotropic or twisted were studied under a polarizing microscope in a birefringent regime of the polarized ordinary and extraordinary light waves [134-143]. Every liquid crystal shows different phase transitional temperatures as follows:

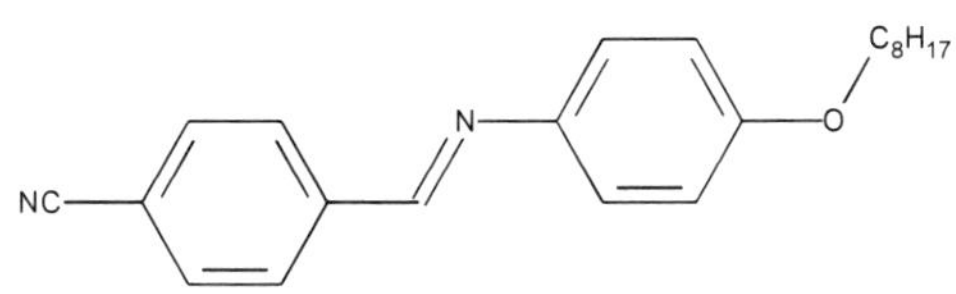

$$CBOOA : I_{109°C} - N_{83,0°C} - S_m A_{73,0°C} - C$$

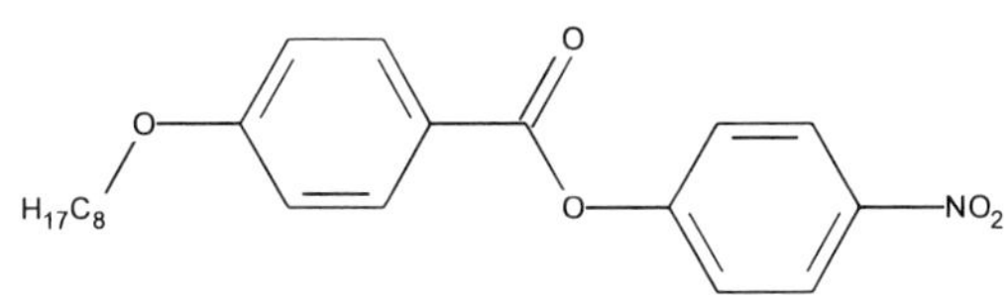

$$NPOB : I_{68°C} - N_{61,5°C} - S_m A_{49°C} - C$$

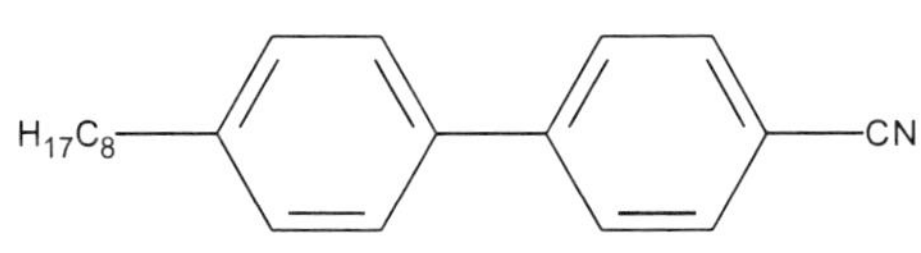

$$8CB : I_{39,5°C} - N_{32°C} - S_m A_{20°C} - C$$

where by I, N, Smectic-A and C are designated the isotropic, nematic, smectic-A and the crystal phases, respectively.

We will study here only the smectic-A phase formed from a deformed nematic (for instance, by the Fréedericks transition) and cooled (in the presence of a voltage) to the smectic-A phase. Finally, the voltage was removed after the formation of the smectic-A phase.

During the performance of such experiments, one should know the value of some material parameters of the liquid crystal studied not only at the nematic-smectic-A phase transition, but also in the entire nematic and smectic-A regions. We cite some papers in this direction as follows: for the liquid crystal CBOOA [144-147], NPOB [148] and for 8CB [149-157].

Thermal transport phenomena in the liquid crystals have been insufficiently investigated both theoretically and experimentally [158-164]. Earlier investigations showed a strong dependence of the thermal conductivity on temperature, liquid crystal orientation and the temperature gradient. The thermal conductivity at the boundary surfaces is usually greater and \there is no dramatic change around the nematic-smectic-A phase transition, i.e., only short-range geometrical factors are important.

Cooling the investigated CBOOA cells clearly results in high temperature gradients in the electrode planes from the lateral metal holders storing the thermal energy in the middle part of the cells. However, it is difficult to understand the temperature distribution along Z (normal to the glass plates). A temperature gradient is expected because of the difference in the thermal properties of the glass on the microscope stage and the air above the liquid crystal cell. This, asymmetry however, is not significant because of the massive metal lateral holders with a homogeneous temperature distribution. On the other hand, the cell thickness (20-30 μm) is much smaller than that (~ 1 cm) of the metal holder. This was confirmed in our experiment by placing one complementary glass plate above the cell. No changes in the experimental results were obtained.

The formation of the smectic-A phase of the liquid crystals was obtained in another type of a liquid crystal cell in which the Teflon spacers were replaced by metallic conductive foils (i.e., we have applied a transversal electric field on a preliminary prepared homeotropic layer) and the two massive metal holders were removed. The liquid crystal cells were heated from below by a thermostatic plate, type “Mettler”, supplied with a temperature regulator. One might expect a temperature gradient along Z, perpendicular to the glass slides of the cells since the upper glass plate of the “Mettler” was open. However, surprisingly we have observed such a temperature gradient only in thick cells with a thickness above 25 μm. This fact was evidenced by the creation of well-defined Cladis-Torza’s stripes. In thin liquid crystal cells, however, the temperature was equal inside the liquid crystal layer which was approximately revealed by typical color scattering properties of the widely used short-pitch cholesterics. For instance, the clear color produced by such a short-pitch cholesteric liquid crystal, 10-20 μm thick, heated from below unambiguously pointed out that the nearly equal temperature over the entire liquid crystal layer – the temperature gradient would lead to observation of iridescent colors. Contrary to this however, a significant temperature gradient exists between the outer sides of the glass plates, confining the liquid crystal, which clearly shows the great importance of the thermal properties of the glass plates depending on their thickness, thermal properties, etc.

Finally, the same liquid crystal cells with replacement of the two metal foils by polymer sheets and using conducting glass plates (an application of a normal electric field) have been used for observation of the smectic-A phase of the liquid crystal 8CB being formed from the strongly deformed twisted nematic regions. The replacement of the SiO orientating layer by

PVA one led to different experimental results which manifests the importance of the surface orientation.

1.3. Microscopic Observations of Various Smectic-A Textures

A normal (vertical) to the glass plates high-frequency electric field was applied across the nematic phase of CBOOA and 8CB confined in cells, above described (for the case of the nematic phase of the liquid crystal NPOB the electric field was with a transversal direction). The frequency of the electric field was between 5 and 20 kHz in order to avoid the hydrodynamic effects and the voltage was between 1.1 U_c (U_c is the threshold voltage leading to the deformation of a well-oriented planar (CBOOA), homeotropic (NPOB) or a twisted nematic layer (8CB)) and 40U_c. The dielectrical properties of CBOOA can be found in References [144-147], of NPOB - in [148], of 8CB in - [149 – [157]. The smectic-A grows in a different manner at cooling, creating various smectic-A textures depending on the value of the voltage applied across the nematic cell. This way of the smectic-A growth is schematically shown in Figure 1 for the case of the liquid crystal CBOOA.

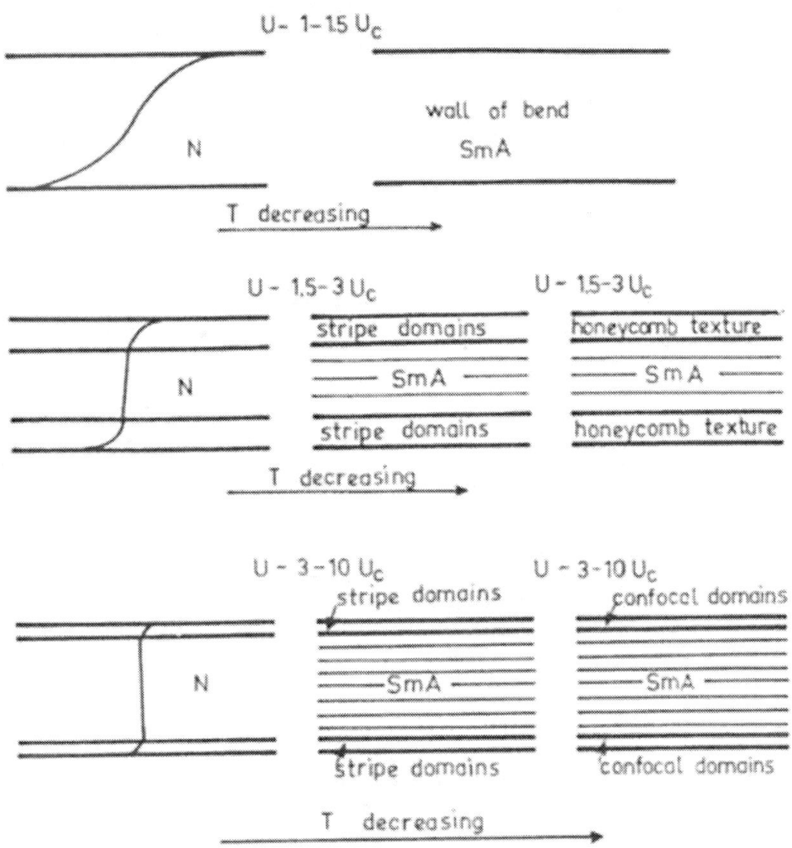

Figure 1.3. Various textures obtained from electrically-deformed nematic phase of initially planar CB00A layer.

Smectic-A bend walls called smectic-A walls of Grandjean were formed from weakly deformed nematic layer when the applied voltage U is in the range of 1-1.5 U_c. Their formation clearly shows that the smectic-A phase grows simultaneously from the whole cell (see Figure 1.3).

Similar walls were observed in the liquid crystal 8CB being formed from initially twisted by $U = 5U_c$ nematic (Figure 1.4).

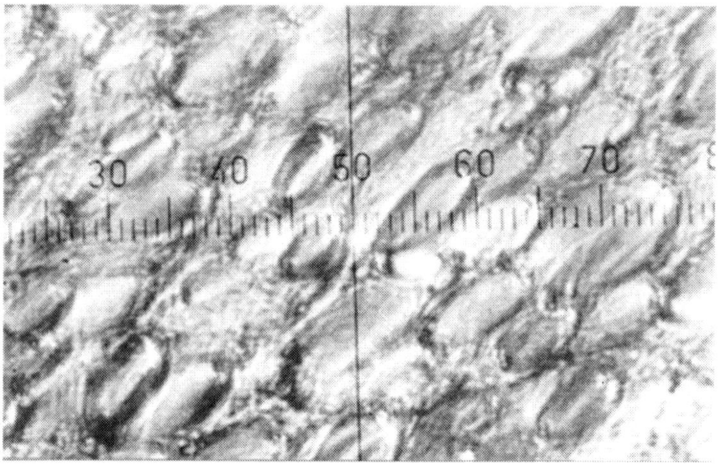

Figure 1.4. Walls of Grandjean in smectic-A.

Many authors have observed such walls. At the increase of the voltage applied across the nematic phase from 1.5 U_c to 3 U_c, and the decrease of the temperature, in the presence of voltage, we have observed two very important effects: the smectic-A phase was formed first in the middle part of the cell, where the phase transitional nematic – smectic-A temperature is higher, and then in the boundary regions where the same temperature according to the theory of de Gennes is lower. In this way, we have confirmed, as early as in the first experiment with the liquid crystal CBOOA, this part in the theory of de Gennes, which claims that the value of the nematic – smectic-A phase transitional temperature decreases when the smectic-A grows from a strongly deformed nematic. For the second time after Cladis-Torza [52] for the case of the liquid crystal CBOOA and for the first time for the liquid crystals NPOB and 8CB we have confirmed qualitatively, by textures, the theoretical results of the de Gennes theory showing that the phase transitional nematic – smectic-A temperature decreases when the smectic-A grows from a deformed nematic. We should mention here that pre-transitional stripes, called striped domains of Cladis-Torza, usually appear in the boundary regions where a strong splay-bend deformation of the nematic exists. According to Cladis and Torza [52], these stripes represent undulation of the phase boundary nematic – smectic-A and definitely show according to Cladis et al., [165] that the nematic-smectic-A phase transition is of a weakly-first order. This fact was confirmed in our experiment (not included here) during the study of the liquid crystal 6O.4 [166]. In order to understand the smectic-A textures observed at application of various voltages across the nematic phase, followed by a free cooling, we should address to some papers devoted to the study of the elastic and plastic behavior of the smectic-A [167-175] for the case of CBOOA. At lower voltage, the stripes of Cladis-Torza transformed quickly either into honeycomb texture (Figure 1.5).

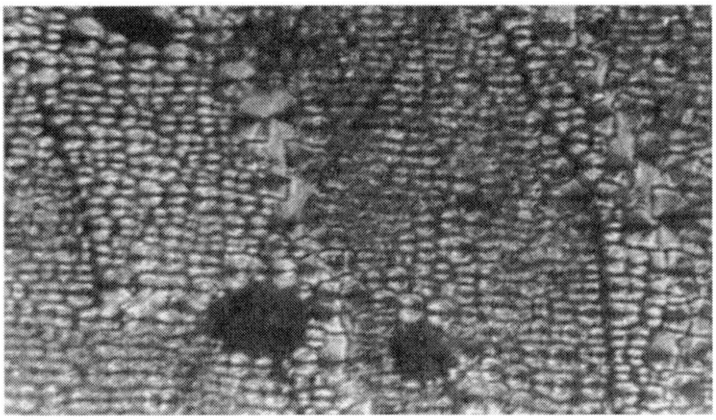

Figure 1.5. Grain boundaries in smectic-A.

or into another kind of a smectic-A texture (Figure 1.6) [176-179],

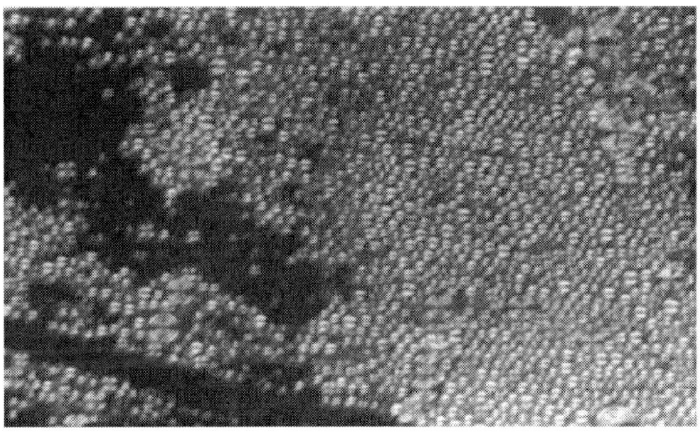

Figure 1.6. Intermediate structures in planar CB00A (26 μm thick, U = 10 $V_{r.m.s.}$, $n_0 \parallel P \perp A$, cooling rate 15°/min). It is observed also circular-line focal conics and homeotropic smectic-A regions with planes parallel to the glass plates. The short side of the photo corresponds to 215 μm.

or into less energetic focal conics {Figures 1.7 and 1.8) [180-188].

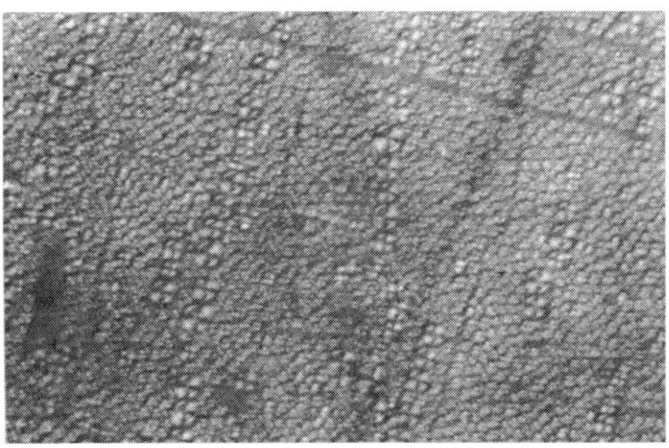

Figure 1.7. Confocal smectic-A domains embedded into a homeotropic smectic-A matrix.

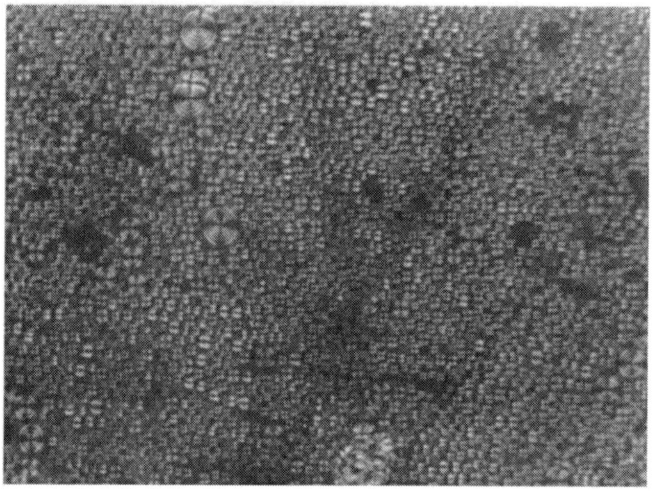

Figure 1.8. Circular-line focal-conics in a planar smectic-A CB00A layer (26 μm thickU=15$V_{r.m.s.}$, n_0 || P ⊥A, cooling rate 15°/min). They are formed in various planes, papallel to the glass plates. The short side of the photo corresponds to 215 μm.

The general type of the focal conics is shown in Figure 1.9.

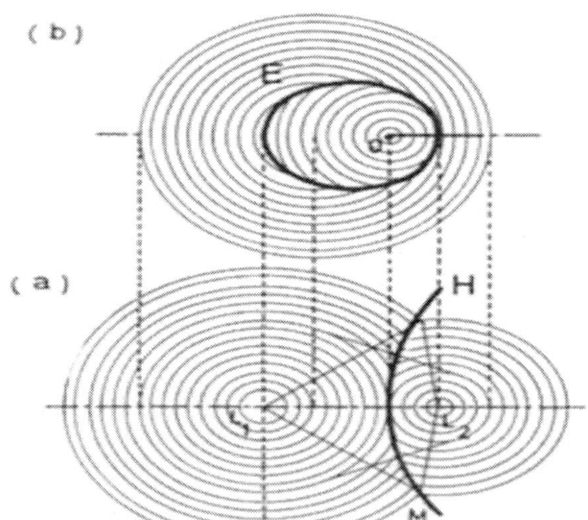

Figure 1.9. Arrangement of the smectic planes according to Friedel's model for the focal-conics. The conjugated conics are drawn in two perpendicular planes (a,b).

Such a type of focal conics was rarely observed by us. In principle, we have observed focal conics in which the ellipse is degenerated into a circle and the hyperbola - into a straight line. The rest part of the liquid crystal is a homeotropic smectic-A with smectic planes parallel to the glass plates confining the liquid crystal under study (see Figure 1.10).

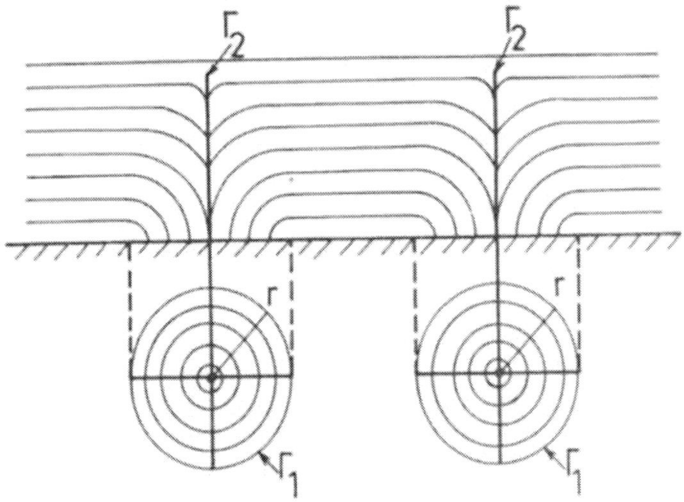

Figure 1.10. Schematic representation of the smectic-A phase obtaned from a strongly deformed nematic phase. The molecules are perpendicular to the smectic layers.

In general, the confocal domains usually appear in different levels inside the liquid crystal cells. They appear only in the boundary regions after the application of a high voltage on the nematic phase. We have observed an intermediate smectic-A texture (Figure 1.11 and Figure 1.12).

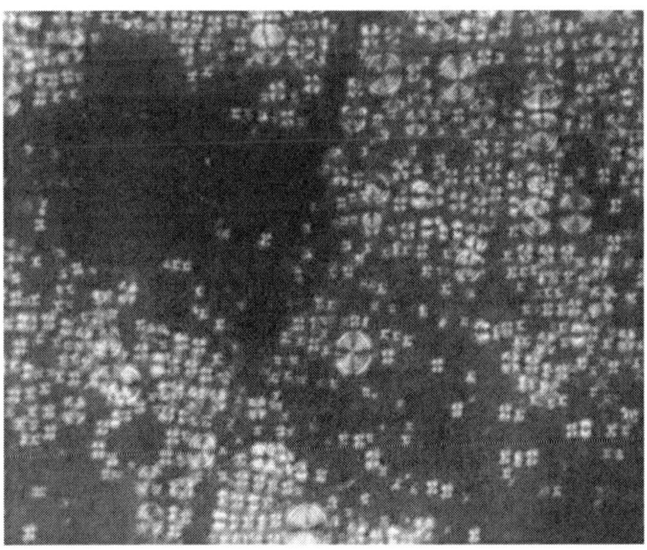

Figure 1.11. Intermediate structures in planar CB00A (26 μm thick, U = 10 $V_{r.m.s.}$, $n_0 \parallel P \perp A$, cooling rate 15°/min). The short side of the photo corresponds to 215 μm.

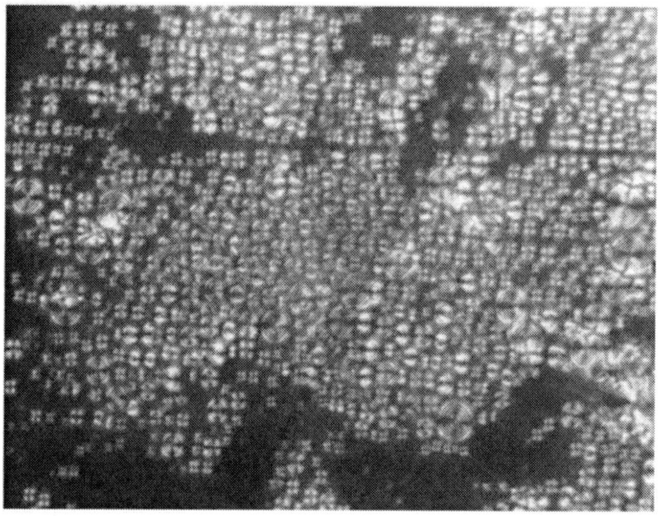

Figure 1.12. Intermediate structures in planar CB00A. The same conditions pointed out in Figure 1.11.

(for the second time after the observations of the researchers from the Thomson-CSF group [189]) representing the formation dynamics of the focal conics after the relaxation of the stripes of Cladis-Torza. However, we have observed the same intermediate smectic-A texture (for the first time) after relaxation of the circle-straight line focal conics into oriented smectic-A except the linear disclinations [69]. These experimental observations permit to conclude: the formation or disappearance of the circle-straight line focal conics at different initial conditions are connected with the intermediate smectic-A textures. In this way, we were able to confirm the claims of Taylor and Kahn [190] and of Pershan and Prost [191] that the impurities play important role at the formation of the circle-straight line focal conics when scattering textures are formed and that the formation of the circle-straight line focal conics is rather heterogenetic (this kind of formation after Chou et al., [192] is connected with irregularities of the glass plates confining the liquid crystal, or with the presence of impurities [193-195] in bulk of the liquid crystal). On the basis of our experimental results, we suggested that the strongly deformed part of the confocal domains is formed firstly which is evidenced under microscope with the formation of the intermediate texture observed as elongated, or ellipsoidal, bright spots. This conclusion is naturally connected with the idea of the heterogenetic growth of the confocal domains. Furthermore, we should focus our attention on another important problem – the filling of the available space of the liquid crystal with focal conics [196-201]. It is interesting to note that the formation of the smectic-A in the boundary regions of our experimental cells is close to the one, predicted by Bouligand [183]. In his model, the ellipses are transformed into circles arranged in the plane of the glass plates; the hyperbolas are transformed into straight lines situated normally to the plane of the glass plates, the plane of the circles respectively. The filling region connected with one pair of focal conics, is a cylinder based on the circle. The filling of the available space of the liquid crystal cells with confocal domains can be effected in different ways: Appolonian packing [196, 199, 200, 201] fractal structures, flexible smectic-A layers. The circle-straight line focal conic domains were generated in our experiment at various levels near the boundaries when the voltage was low, at one level when the voltage was high, and had a radius between 1.3 and

2.3 microns. Confocal domains with smaller dimensions were not generated due to energetic causes.

All photographs were taken with natural light between crossed nicols. We note that this position of the nicols is more favourable for optical investigations of the smectic-A structures observed. The concentric rings observed in the domains result from changes in the total phase difference between ordinary and extraordinary wave (as noted by Perez and Proust [202]) and are seen only for splay orientations between crossed nicols. Similar optical pictures (well described by Niessen and Ouden [203]); (see also [204]) were illuminated by a convergent light beam). In connection with these pictures, it is interesting to note their local optical biaxiality which has been investigated by Galerne [205] in oriented defects in samples of lamellar liquid crystals. The circular shape of the confocal domains may also be observed without nicols in natural light [202, 206]. These shapes are clearly distinguished from the polygon shaped picture obtained under the same optical conditions [202,206].

In conclusions, favorable conditions for observing the confocal domains generated in the boundary regions are the use of natural light, crossed nicols and high magnification (x 400). The formation of differently distributed confocal domains is more easily obtained by cooling from deformed nematic layer than by using special means for mechanical undulations of the smectic-A layers [207].

NPOB liquid crystal was studied in liquid crystal cells with a thickness of 10 μm excited by a transversal electric field applied on two folia bands which were posed at a distance from 0.65 mm to 2.5 mm. The initial orientation of the nematic phase was either homeotropic, achieved after a careful cleaning of the glass plates, or after use of solution of Lecithin (several weight. percent in ethanol) or planar orientation achieved by evaporation of SiO on the glass plates under vacuum evaporation. The dielectric properties of NPOB, as mentioned above, have been studied in Reference [148], Barnik et al., have determined the value of the dielectric anisotropy in the nematic and smectic-A phase to be around + 15. In addition, the elastic and viscosity coefficients are also known [148].

An AC electric field with a strength between E_{th} (the threshold electric field was in the range of 200-250 $V_{r.m.s.}$/cm) and 35 E_{th} was applied across the liquid crystal films under study in both the nematic and smectic-A phase in order to obtain a complete orientation of the nematic molecules in the middle part of the samples being parallel to the electric field direction (positive value of the dielectric anisotropy) and two strongly deformed liquid crystal nematic regions at the two glass slides(one strongly deformed region in the hybrid-aligned cells). The frequency of the electric field (f = 5 kHz) was sufficiently high to ensure the orienting action of the dielectric torques without any hydrodynamic movement of the fluid. The excitating voltage was switched off after the formation of the smectic-A structures.

The matching of the boundary conditions with the already formed smectic-A in the middle bulk regions of the liquid crystal cell for the case of NPOB with a homeotropic orientation and a transversal electric field proceeded in quite a different way (Figure 1.15).

Let us stress, that the formation of the smectic-A near aluminium foils is quite different, as can be seen from Figure 1.16 (for details see the text under the Figures 1.13 and 1.14):

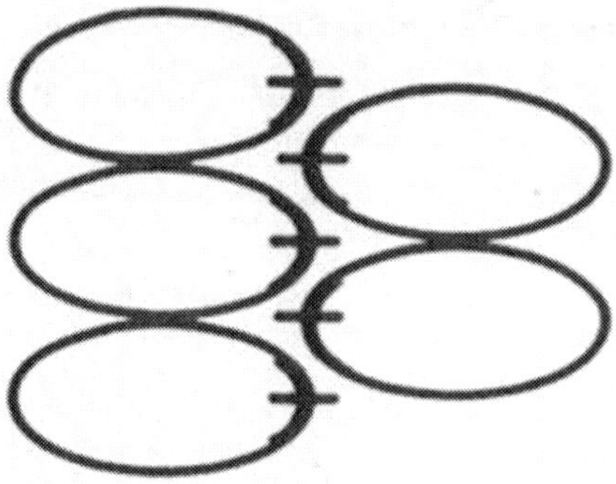

Figure 1.13. Schematic representation of the projection of the ellipses on the glass plate.

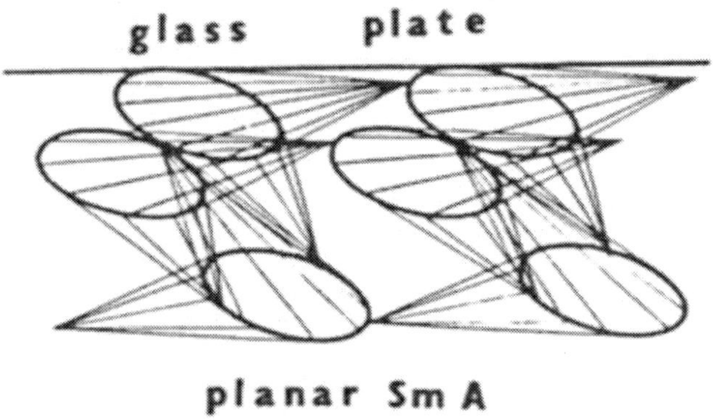

Figure 1.14. Marignan-Malet-Parodi association of confocal pairs of ellipse-hiperbola. A view from the lower side in planar smectic-A being formed.

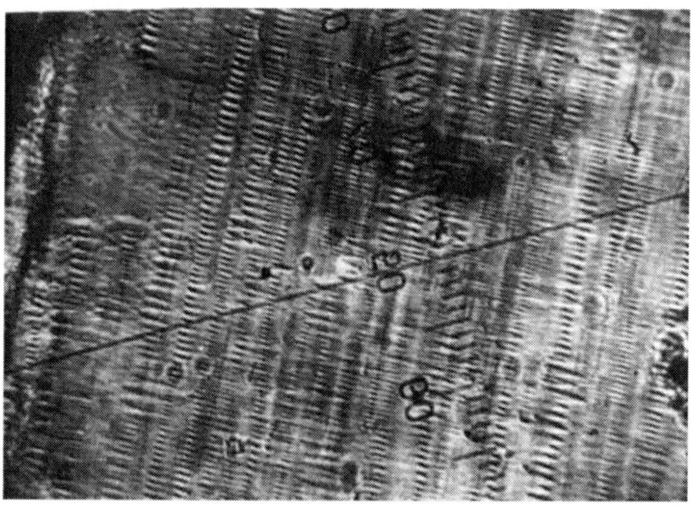

Figure 1.15. Confocals' view of regular arrays observed in NPOB smectic-A film 10 μm thick, U=700 $V_{r.m.s.}$, f=5 KHz(l=1,5 mm), initially crossed nicols, the electric field is applied along the bisectrice of the nicols, the polarizer is then rotated at an angle of 40°, the long side of the photo corresponds to 230 μm.

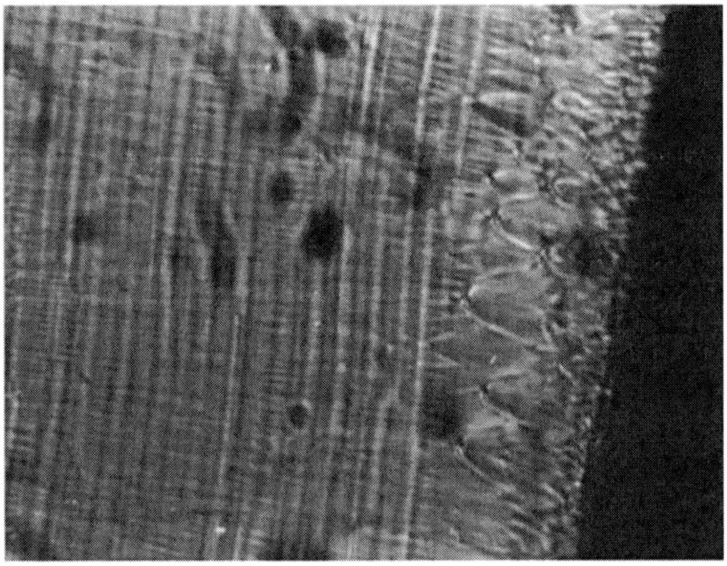

Figure 1.16. View of smectic-A NPOB layer in the vicinity of the aluminium spacer (d=10 μm, U=300$V_{r.m.s.}$, f=5 KHz, l=2,5 mm), crossed nicols, the electric field is applied along the bisectrice of the nicols, the long side of the photo corresponds to 230 μm.

Similar arrays of confocal domains have been observed in References [208-210]. The construction laws for the confocal domains had been clarified for the first time by Grandjean [211] and Friedel [212]. All these cases have been considered in detail by Bouligand [183] as well as in the References [213] and [214].

The Marignan-Malet-Parodi association of the confocal ellipses-hyperbolae is shown in Figure 1.14. This arrangement actually consists of two opposite Grandjean's walls with ellipses and hyperbolas situated in such a way that two of the generatrices coincide (i.e., there is one common generatrix) and the other ones must be either normal to the outer smectic-A layers where the molecules are parallel to the direction of the electric field or must be parallel to the glass plates.

In the model of Marignan et al., the ellipses can cross each other since they are very elongated and their distorted parts are very small and located in opposite places (see Figure 1.13).

On the other hand, this special arrangement of the confocal domains permits the easy adaption of the normal smectic-A layers which are in the middle part of the samples (planar position of the smectic-A molecules which are parallel to the electric field) with the tilted position of the smectic-A layers in the boundary regions (see Figure 1.14). The different sizes and eccentricity of the ellipses are probably due to the non-equal surface anchoring in the various regions of the liquid crystal areas and less probably to growing causes.

Our experimental results show that the two arrays of confocal domains cross each other in such a way that the bright lines are created due to focuses of the ellipses, and the dark lines are obtained from the overlapping of the ellipses situated on different levels of the liquid crystal. Here we must mention that the filling of the rest available space can be performed either by formation of planar smectic-A layer, when the surface energy of the smectic-A is overcome, or by flexible smectic-A layers.

In hybrid-aligned liquid crystal cells, edge dislocations [215-233] were formed at the glass plate determining planar orientation and arrays of confocal ellipse-hyperbola were formed at the glass plate determining a homeotropic orientation. This case is shown in Figure 1.17.

The increase of the voltage up to 500 $V_{r.m.s.}$ (the distance between the aluminum foils was in the range of 1 mm, the thickness of the cell 10 μm, f = 5 kHz) and the rotation of the cell at an angle of 45^0 between crossed nicols led to the observation of a very rare texture shown in Figure 1.18.

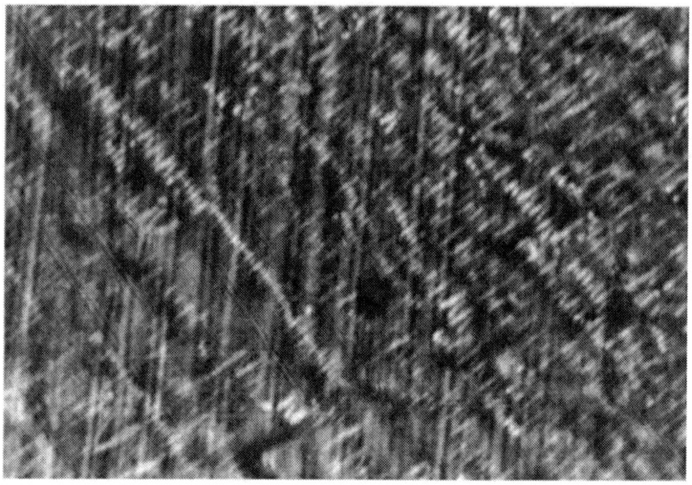

Figure 1.17. A picture showing the simultaneous existence of aligned arrays of focal-conics formed at one of the boundary and smectic-A edge dislocations formed at the other boundary (d=10 μm, U=150$V_{r.m.s.}$, f=5 KHz, l=1 mm), crossed nicols; the electric field is applied along the polarizer, the long side of the photo corresponds to 230 μm.

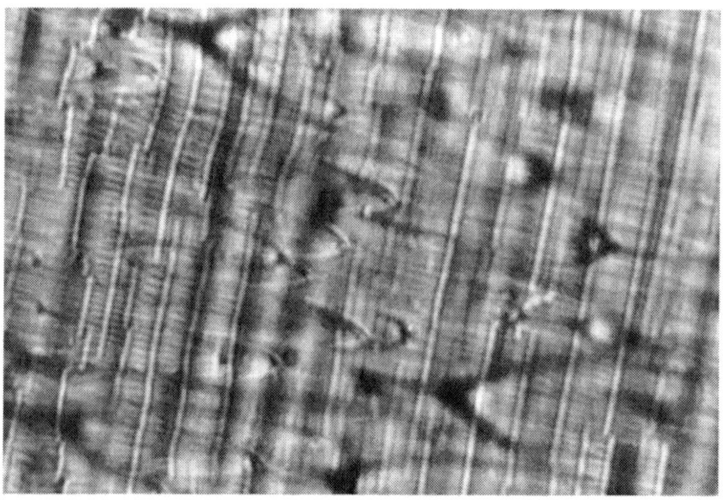

Figure 1.18. A combined picture of overlapping confocal domains being formed in hybrid-aligned homeotropic-planar NPOB smectic-A film (d=10 μm, U=500$V_{r.m.s.}$, f=5 KHz, l=1 mm) crossed nicols, the electric field is applied along the bisectrice of the nicols, the long side of the photo corresponds to 230 μm.

It represents a superposition of two radically different confocal textures: arrays of pair confocal ellipse-hyperbola which were observed and explained for the first time by Marignan et al., and Parabolic focal conics observed and explained for the first time by Rosenblatt et al., [234] and studied by other authors [235-241]. The images shown in Figure 1.17 and Figure 1.18 clearly show that the smectic-A phase has appeared first in the middle part of the liquid crystal cells and after that in the boundary regions, which unambiguously confirms the theory of de Gennes for the nematic – smectic-A phase transition.

It is important to note that arrays of confocal ellipse-hyperbola can form directly from the nematic phase by cooling and growing the smectic-A phase without the formation of intermediate edge dislocations or $\pm\pi$ disclinations. Inversely, at heating, the ends of the ellipses carrying out the smallest elastic energy disappear first, followed by the more energetic focal regions with strongly curved smectic-A layers. Consequently, the observed impressive arrays of focal conics are highly organized smectic-A textures with special arrangement possessing much less elastic energy as compared to the edge dislocations or the various kinds of grain boundaries.

1.4. Filling up the Space with Focal Conics: Discussion

The formation of the smectic-A with arrays of confocal domains can be correctly interpreted if one knows the possible causes for their growth. According to Williams [242] such arrays of confocal domains (whole or half Grandjean walls with parallel hyperbolas and arrays of ellipses with equal eccentricity touching their heads and ends) can appear due to the process of competition between surface and bulk forces under the influence of external electric fields. In order to answer the question how the filling of the available space with confocal domains occurs, it is necessary to know the three main laws for the possible association of the focal conics: the law of non-penetration of the focal conics, the law of possible common generatrics of the focal conics and the law of the corresponding cones. In accordance with these laws and the possible surface orientation of the smectic-A director [243-246], the ellipses can be vertical, inclined or parallel to the plane of the glass plates confining the liquid crystal (the position of the hyberbolas can be easily determined in accordance with the construction of Dupen cyclides, i.e., they must pass through the foci of the ellipses and must be in planes, perpendicular to the plane of the ellipses). Further, the ellipses can be situated on the glass plates or in depth of the liquid crystal layer. They can be arranged in straight arrays, when the ellipses are equal, touching their heads and ends or can be arranged in "treilles" when the eccentricities of the ellipses are different. Non-parallel ellipses lying in one plane can organize themselves in polygonal textures, etc.

1.5. Possible Physic Causes Leading to the Formation of Circular-Straight Line Focal Conics and Regular Arrays of Focal Conics Ellipse – Hyperbola in the Smectic-A Phases

The formation of various kinds of circular-straight line focal conics from a planarly oriented smectic-A is determined, first, by their possible building (see the results of Friedel and Lavrentovich-Kleman) and second, by their heterogenic formation. It is clear that they are formed due to energetic causes and it was evident that when the surface energy is larger, then the confocals are smaller. The value of the surface energy of the liquid crystal and the value of the surface tilt of the director are of crucial significance for the kind of the smectic-A textures formed. For example, arrays of "grain boundaries" can be formed in planar and well-aligned smectic-A layers with a large surface energy. The "grain boundaries" and the dislocations, however, cannot be formed in the case of a homeotropic orientation of the smectic-A with smaller surface energy since all these smectic-A textures cannot be supported by the boundaries. This is confirmed by the repeatable creation of the arrays of focal conics for different thickness of the liquid crystal films being changed between several micrometers and 50 μm and consequently for different temperature distribution. On the other hand, these smectic-A textures crucially depend on the liquid crystal surface orientation and anchoring as revealed by the repeatable appearance of the arrays of confocal domains in the same regions of the liquid crystal cells. As claimed by Williams [242], the simple bending of the smectic-A planes leading to the formation of grain boundaries is needed from a surface energy in the range of 4×10^{-2} erg/cm^2 which is higher in comparison with the value of the surface energy, calculated for the case of homeotropic orientation of the smectic-A.

1.6. Determination of the Surface Energy of the Nematic and Smectic-A Phases

a) Determination of the θ-Surface Energy of the Nematic Phase of CBOOA

The application of a strong electric field across a nematic with positive dielectric anisotropy and a finite surface energy must lead to a complete orientation of the director along it. It is clear that this complete alignment is difficult to observe for a liquid crystal, which possesses only a nematic phase. This task, however, can be considerably simplified when smectic-A phase follows after the nematic phase. For instance, on decreasing the temperature, aligned smectic-A can be obtained directly from the nematic under the action of a high electric field (without formation of the Cladis-Torza stripes), when the coherent length $(4\pi K/\Delta\varepsilon)(1/E)$, showing the penetration depth of the splay-bend deformations (K is the mean elastic constant and $\Delta\varepsilon$ is the dielectric anisotropy) tends to zero [13]. The value of the constant W_{SN} showing the strength of the surface anchoring can be calculated from the balance between the negative maximal electric energy (per cm^2) of the liquid crystal and the positive maximal surface energy of the same area:

$$\left(|\Delta\varepsilon| / 8\pi\right) E^2 d = W_{SN}(\pi / 2) \tag{1.4}$$

We have calculated this surface energy for the case of CBOOA(U = 30 V, d= 26 μm) and obtained the following value:

$$W_{SN} \cong 0{,}85\ erg / cm^2 \tag{1.5}$$

for the glass plates covered by *SiO* under vacuum evaporation.

b) Estimation of the Surface Energy of the Smectic-A Phase

Meirovich and Freed [247] considered for the first time the problem concerning the surface energy of the smectic-A phase [248-258]. They made very important remarks about the penetration of the smectic torques in depth of the smectic-A cells. According to these authors one and the same substrate can create a strong anchoring of the nematic director and a weak anchoring of the smectic-A director. This difference is naturally related to the peculiar character of the nematic and smectic-A phases. They concluded that the strength of the surface anchoring of a homeotropic smectic-A phase depends on a dimensionless ratio $(\pi/2)K_N/(\overline{B}\,K_1)^{1/2}$ where K_N is with a dimension of force/distance [dyn/cm](it is usually designated by the letters W_s or C and has a dimension of erg/cm^2 and ensures the homeotropic orientation of the smectic-A phase), $\overline{B}$ shows the dilation or contraction of the smectic-A layers and K_1 is the elastic constant of the splay showing the splaying of the smectic-A director inside the smectic-A layer. Comparing this term with the extrapolation length of the nematic $(K_N L)/(\pi K_1)$ introduced for the first time by de Gennes, Meirovitch and Freed [247] have concluded that the surface anchoring of the smectic-A is softer relative to that of the nematic, and that the smectic-A deformation can easily be realized and penetrate in depth of the liquid crystal cells under study. Poniewierski and Holyst [251] suggested a general kind of the surface energy of the smectic-A, including a vertical displacement of the smectic-A layer and the surface tension of one free smectic-A layer. Cagnon and Durand [169], independently and simultaneously, have considered the very important problem related to the positional anchoring of the smectic-A layers. Lavrentovich et al., [252, 253] have estimated the value of the surface energy of the smectic-A when it is formed with the aid of dislocations and found a very big surface energy (0.3-30) erg/cm^2.

b1) Determination of the θ-Surface Energy of the Smectic-A Phase of CBOOA

This surface energy can be estimated from the theoretical results obtained by Kléman [260-263] for the elastic energy of the circular - straight line confocal domain of the smectic-A:

$$W_{cd} = 2\pi^2 K_1 r_c [\ln(2r_c / r_{cor}) - 2] \quad (1.6)$$

where r_c is the radius of the circle, r_{cor} is the core radius [259] and K_1 is the splay elastic constant.

The energy of the smectic-A in the surface regions is maximized and takes the value W_s ($\pi/2$) when the smectic layers are planar and parallel to the plane of the Electrodes. This total surface energy is equal to the sum of one half of the energy of the confocal domains just above their disappearance:

$$W_{cd} = \pi^2 K_1 r_c [\ln(2r_c / r_{cor}) - 2] \quad (1.7)$$

and the energy of the core line:

$$W_{cl} \cong 2E_{cor} r_c \quad (1.8)$$

The energy of the core line per unit length is of the order of K_1. The core radius of the circle disclination line as well as of the straight line is of the order of thickness of several smectic layers ~ 100 Å. Then for glass plates treated with SiO, for the lowest value for r_c obtained in our experiment which is of the order of 1.3 μm, we have established the mean value of the θ-surface energy for the case of the smectic-A:

$$W_s = (\pi K_1 / r_c)\,[\ln(2r_c / r_{cor}) - 2] + (2\pi K_1 / r_c) = 0{,}14\, erg / cm^2 \quad (1.9)$$

The comparison of the magnitude of the elastic energies for the case of both the nematic and smectic-A cases show that the θ-surface energy for the case of the nematic phase is 6 times bigger. The large difference between the θ-surface energy of the nematic and smectic-A phases is due mostly to the existence and type of the very confocal domains in the smectic-A phase. First, it is well known that confocal domains of submicron and nanosize have not been observed. Kléman has clarified this problem in 1976 [185]. Second, Fournier [256] showed that the first smectic-A layer surrounding the linear disclination should be compressed, which increases considerably the elastic energy of the confocal domains. This claim is able to explain also the non-observance of confocal domains with sub-micron sizes (see also the papers [252] and [253]). Independently of these causes, it is clear that the estimation of the energy of the confocal domains is approximate, the accuracy decreases when their size decreases and, inversely, increases when their size increases. It is also clear that the accuracy of the estimation of the surface energy of the smectic-A increases when the magnitude of the surface energy decreases. For example, confocal domains cannot be formed for a surface determining a θ-surface energy smaller than 10^{-2} erg/cm^2, a result which is very important for the practice. Hataevich and Zeinally [264] have proved experimentally this suggestion (the value of the polar θ-surface energy of the smectic-A phase can be found in the papers [187, 244]).

b2) Estimation of the Total Surface Energy of the Smectic-A Layer for POB

For the first time, in 1983, we have estimated the surface energy of the smectic-A phase for the case of homeotropic anchoring. By choosing two pairs of well-formed ellipses [68] and using the elastic energy of confocals obtained by Kléman, we have calculated the order of the total θ-surface energy:

$$W_s \int_0^{\theta_0} f\left(\sin^2\theta\right) d\theta \sim 1.56x10^{-2} erg/cm^2 \tag{1.10}$$

It is interesting to compare this energy with that calculated from the exact formula of Kléman and Lavrentovich [263](see also [265]):

$$W_{cd} = 4\pi a\left(1-e^2\right)K\left(e^2\right)\left[K_{11}\ln\frac{2b}{r_{cor}} - \left(2K_{11}+\overline{K}\right)\right] \tag{1.11}$$

In this formula, a is the big semiaxis of the ellipse, e is the eccentricity of the ellipse, b is the short semiaxis, $K(e^2)$ is ellipitic integral of the first kind. The total surface energy is calculated according to the same formula used in [68]:

$$W_s \int_0^{\theta_0} f\left(\sin^2\theta\right) d\theta = \frac{l}{mn}\left\{W_{cd} + 2E_{cor}r\right\} \tag{1.12}$$

with the following numerical values extracted from the marked region: $a = 5\ \mu m$, $b = 3\ \mu m$, $e = 4$, $r_c = 100\ \AA$, $r = 24\ \mu m$, $E_{cor} = K_1$, $m = 23\ \mu m$, $n = 6\ \mu m$, $l = 4$, $\overline{K} = 0$ [68].

The value of the surface energy is calculated from the above pointed out two formulae with accuracy to the second sign after the decimal point as follows:

$$W_s \int_0^{\theta_0} f\left(\sin^2\theta\right) d\theta \cong 3,56x10^{-2} erg/cm^2 \tag{1.13}$$

It is clear that this energy cannot support the formation of dislocations in the liquid crystal cell. In addition, it is clear that the approximate value of the surface energy is about two times smaller than the real value.

1.7. Determination of the Maximal Tilt Angle of the Ellipses with Marignan- Malet-Parodi Arrangement

The maximal tilt angle of the ellipses, a very interesting quantity, can also be calculated according to the formula given by Williams [242]:

$$\theta_0 = 2\theta = 2arctg\left[\left(4a^2/4b^2\right)-1\right]^{1/2} \tag{1.14}$$

It is varying from 72^0 for the chosen pairs of confocal ellipse-hyperbola, used in our calculations, to $10\text{-}20^0$ for the more elongated ellipses. On the other hand, it is clear that the surface tilt of the smectic-A director varies between the above-cited values and 0 for planar orientation of the smectic-A director along the generatrix of the hyperbolas touching the glass plates.

1.8. Discussion on the Parabolic Focal-Conics Appearance

Parabolic focal conics were observed in the liquid crystals NPOB. They appear as "black pitchfork" of various size and two opposite orientations (Figure 1.18). According to the determination of Rosenblatt et al., [234], the parabolic-focal conics are the limit in the form of the focal-conics and represent two parabolic lines, each passing through the focus of the other parabola, and they are situated in two mutually orthogonal planes. Rout and Choudhary [266] on the basis of theoretical results obtained by Kléman [261], have discussed in detail the elastic energy of the dislocation lines and the Parabolic focal conics. These authors have derived the following approximate formula for the energy of one parabolic focal-conic:

$$E \sim \frac{\pi K_{11}}{8} f\left(\frac{R}{f}\right)^4 \ln\left(\frac{R^2}{4fr_c}\right) \quad for\ R >> f \tag{1.15}$$

In this formula K_{11}, f, r_c and R are correspondingly the elastic constant of splay, the focal length of the parabola, the radius of the core of the parabolic disclination line and the radius of the circle, surrounding the parabola within the frame (the boundaries) of the sample. It can be represented in the more convenient form:

$$E \sim \frac{\pi K_{11}}{8} R\left(\frac{R}{f}\right)^3 \ln\left(f\frac{R^2}{4f^2 r_c}\right) \quad for\ R >> f \tag{1.16}$$

The elastic energy increases with R and tends to infinity when the focal length f tends to zero (R/f is constant). In this way, these parabolic focal-conics domains are not energetically advantageous. However, when $3f \geq R$ they can have smaller elastic energy. It has been proved that the elastic energy of the Parabolic focal conics is by two orders of magnitude smaller than that of the dislocation lines.

1.9. General Discussion on the Method Proposed for Growing the Smectic-A Phase

It is clear that scattering smectic- A textures with storage can be obtained in two waysafter application of AC electric field across the smectic-A phase [267-271] or by the method here considered. The comparison of these two methods clearly shows that in our method the necessary voltage is around 4V whereas the necessary voltage, applied directly to the smectic-A, should be around 20 V. This big difference is due to the radically different mechanisms leading to the formation of smectic-A textures. In our method, the electric field was applied across the preceding nematic phase in order to deform the nematic in the surface regions of the liquid crystal cells while in the second way, the electric field is applied directly to the smectic-A in order to create deformations of the type, suggested by Parodi [176], Very often, the deformation in the smectic-A phase is more complex, requiring big values of the electric field and leading to a deformation of the smectic-A phase not only in the bulk, but also in the surface regions [252, 253]. In these cases one notes textures very similar to the cholesteric "oily streaks". Further, arrays of confocal domains are formed inside such "oily streaks". They can also be obtained at mechanical stresses applied to the smectic-A films [256]. It is obvious that such methods [256] cannot be applied in the usual smectic-A displays [272]. The suggested by us new method of obtaining high-quality scattering smectic-A textures has two extra advantages: first, these scattering smectic-A textures do not depend on the temperature in the smectic-A phase and second, they do not depend on the thickness of the liquid crystal cells. Inversely, the obtained smectic-A textures via application of AC electric field directly on the smectic-A phase, are strongly temperature-dependent. Furthermore, they are strongly dependent on the thickness of the liquid crystal cells [252, 253, 267-271, 273]. The concomitances of our method are similar to those appearing in the laser-addresed displays (see below).

The second important problem which should be mentioned is connected with the elastic energy of the confocal domains and the surface energy of the smectic-A phase. All our experimental results clearly show that the creation of confocal domains in the boundary regions near the glass plates is supported by relative small energies in the range of 10^{-2} erg/cm^2. Evidently smaller surface energy cannot support the formation of confocal domains [264]. Another important question is "where" and "how" the smectic-A phase arises from the nematic. Our experimental results clearly show that the domains of Cladis-Torza can be suppressed after application of a sufficiently high voltage across the nematic phase. On cooling, (in the presence of this high voltage), smectic-A phase grows continuously from the nematic phase of the liquid crystal with nematic-smectic-A phase transition, slightly first order or nearly second order. The existence of various disclinations plays also a crucial role [55]. Such a formation of the smectic-A phase was observed by us from a previously-twisted nematic phase (see below).

1.10. The New Electrooptic Effect and Possibilities for Application in the Laser-Addressed Liquid Crystal Displays

Kahn [274] discovered the laser-addressed displays (see also References [275-298]). In that display, the smectic-A layer with a homeotropic orientation of the director is locally heated into the isotropic phase by the thermal energy of an infrared laser beam. When the heating is suppressed, the liquid crystal cools rapidly to the initial temperature (often the room temperature) and goes back to the smectic-A phase which is not transparent since it contains strongly light scattering smectic-A centres retained for a long time. The "erasure" of the scattering structures is performed by the application of a high-frequency electric field assisting the thermal heating-cooling process. Chou et al., [192] have discussed a number of important questions connected with the scattering of light by confocal domains such as the significance of the surface energy of interaction between the nematic and smectic-A phases as well as the influence of the surface tilt of the director. According to their results, the confocal domains generated in the surface regions are more stable than those formed in the bulk of the liquid crystal and their density depends crucially on the transition enthalpy and the value of the surface tilt angle of the director of both the nematic and smectic-A phases. These authors show also the great importance of the treatment of the glass plates for the growing of the smectic-A phase. It is clear, however, that the presence of the glass plates only cannot ensure the creation of homogeneous scattering smectic-A centers. The liquid crystal cells are not convenient for applications in thermally-addressed liquid crystal displays when the glass plates are not treated or are treated in such a way, as to ensure weak surface anchoring. Such scattering centers can be created only if the formation of the smectic-A textures is controlled by surface forces and by energy balances. Our presentation about the creation of smectic-A scattering centers makes it clear that we are able to improve the cell suggested by Kahn. Here we should stress that our method permits the creation of confocal domains in the surface regions with a maximal homogeneity as well as with a maximal density of the scattering centers. We should stress also that in contrast to the results of Chou et al., [192], the value of the voltage applied across the smectic-A in our case, first, can be changed in a large interval without influence on the formed confocal domains (they disappear with a threshold at a much higher voltage) and second, the quality of the formed smectic-A domains cannot be changed by the rate of the temperature change. We show here as an example a possible scheme for the application of our cell in the laser-addressed smectic-A displays (Figure 1.19).

The initial transparent homeotropic smectic-A state, designated here by (1), can be obtained through the application on the nematic phase of a sufficiently high a.c. voltage which must be above $30U_c$ for the case of the liquid crystal CBOOA, where U_c is the Freédericksz threshold voltage for a planar nematic layer, followed by cooling the nematic until the appearance of the smectic-A phase. The second step is the decrease of the voltage between $3U_c$ and $10U_c$ followed by the local heating of the smectic by infrared laser beam, or by other heat sources, to the isotropic phase. This is revealed by the position (2) in our scheme. After the removal of the heating source, the liquid crystal goes back to the smectic-A phase. The scattering smectic-A centers are created in the presence of a voltage, which after that is removed (position (3) in our scheme). The "erasure" or removal of the scattering smectic-A centers can be local or total and in each of these cases it is a highly a.c. voltage-assisted thermal process. In the presence of a high a.c. voltage, orienting the whole nematic phase, the scattering regions of the cell are initially heated by an infrared laser beam up to the isotropic phase (positions (4) and (5) in Figure 1.19). One stops the heating and the voltage is removed when the cell has returned to the initial homeotropic orientation of the smectic-A phase (position (6) in Figure 1.19).

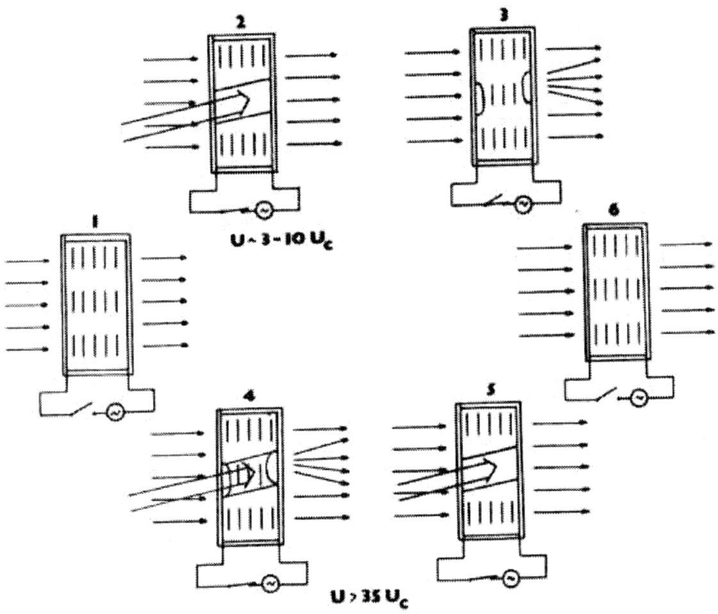

Figure 1.19. Operation of the thermally-addressed display, proposed by us, with stating figures;"writing" cycle (1)-(2)-(3); "erasing" cycle (4)-(5)-(6).

Scattering centers in the smectic-A phase can be obtained by electro-hydrodynamic effects as well and they found application in the smectic-A displays with storage. We note here that these two methods are not antagonistic. Something more, they can be combined if the low-frequency and high-frequency voltages are applied simultaneously and one uses a thermic cycle leading to the appearance of the following phases: nematic – smectic-A – nematic. We should, however, stress here, that the nematic – smectic-A phase transition is influenced by the hydrodynamics and this will complicate the application of our method.

Finally, one should note that the application of an electric field on a liquid crystal cell with a temperature gradient has led to the discovery of new laser-addressed displays [295] and to a more detailed study of the phenomena in the electro-thermo-addressed smectic-A displays [299-301].

We will focus our attention now on two problems: distribution of the temperature in the laser-addressed displays [302-306] and the creation of the scattering centers in the smectic-A. As we have noted above, the first question is more important for the originally suggested by Kahn display and is less important for our cells. For instance, using a Mettler hot storage and a thermoregulator, we have checked different temperature variations from 0.1^0C/min to 10^0C/min as well as a free cooling (~ 15^0 C/min). We have obtained in all cases similar experimental results. This is due to the fact that we have used a liquid crystal with nematic-smectic-A phase transition which is close to second order. In all these liquid crystals, the formed textures are confocal domains. The formation of these domains depends on a number of factors such as the initial orientation of the nematic, its surface orientation, the value of the applied voltage, etc. Nevertheless, it is always possible to find an appropriate liquid crystals with appropriate boundary conditions which can work by the above-pointed scheme. The second problem we wish to clarify is related to the scattering properties of the various packed confocal domains [307-309]. The obtained experimental results show that the scattering properties strongly depend on the kind of the scattering centers. The scattering angle is relatively small (10-15^0) when the confocal domains are randomly distributed. In addition, diffraction effects appear when the confocal domains are more regularly arranged.

1.11. A Reversible Electro-Optic Effect in the Smectic-A

It is well-known that reversible electro-optic effects in the smectics can be rarely observed. For example, de Gennes [13] has shown that optic changes in the structure of the smectic-A phase are very difficult to observe even at very high voltage excitation. In effect, he showed that such changes (transitions) in the smectics-A will be very weak and optically non-observable (the so-called “ghost” transitions). Parodi was the first author who showed that it is possible to observe experimentally “grain boundaries” at appropriate high voltage [176]. Indeed, such “grain boundaries” have been observed by Goscianski et al., [177]. The early experimental work of Tani [310] as well as the experimental work of Le Berre and co-authors from Thomson-CSF [189] have stimulated further efforts of the researchers for application of the smectic-A phase in the displays. Raynes [286] also made useful remarks for the application of high-frequency electric field on the smectic-A phase. In effect, the planar-to-homeotropic smectic-A transition found a realization in the liquid crystal displays with storage. In short, up to 1988 only one specific reversible effect in the smectic-A has been discovered which was used in the liquid crystal displays with storage. It was necessary to discover other reversible electro-optic effects in the smectics-A too [311].

Such an electro-optic effect was obseved in the smectic-A phase of the liquid crystal 8CB grown from a $\pi/2$ twisted nematic phase in the vicinity of a second order nematic-smectic-A phase transition (in a 5^0 C temperature region). The same experimental results can be obtained if one uses a smectic-A with preceding large-pitch cholesteric phase [69]. The solution of this problem shows that such an effect can exist in a wider temperature interval if the formation of the smectic-A is heterogeneous and if it is connected with the suitably treated boundaries of the liquid crystal cells [312]. As noted in the beginning of this chapter, this liquid crystal is one of the most frequently studied with nearly second order nematic-smectic-A phase transition and a very small latent heat at this transition: 0.304 kJ/mol [69]. The dielectric anisotropy is positive and equal to 8.2 at room temperature (T = 22^0 C) [267]. The initially twisted by $\pi/2$ 8CB nematic layer was subjected to a voltage with a frequency 19 kHz and an amplitude 10 $V_{r.m.s.}$ and cooled down to the smectic-A in the presence of the voltage. The glass surfaces were treated by SiO, evaporated at an angle of 30^0 to the surface. The threshold voltage was measured to be about 1.6 $V_{r.m.s.}$. At the nematic-smectic-A phase transition, the usual pretransitional stripes observed by Cladis-Torza [52] were not formed. Instead, we have observed textures shown in Figure 1.20, a and b.

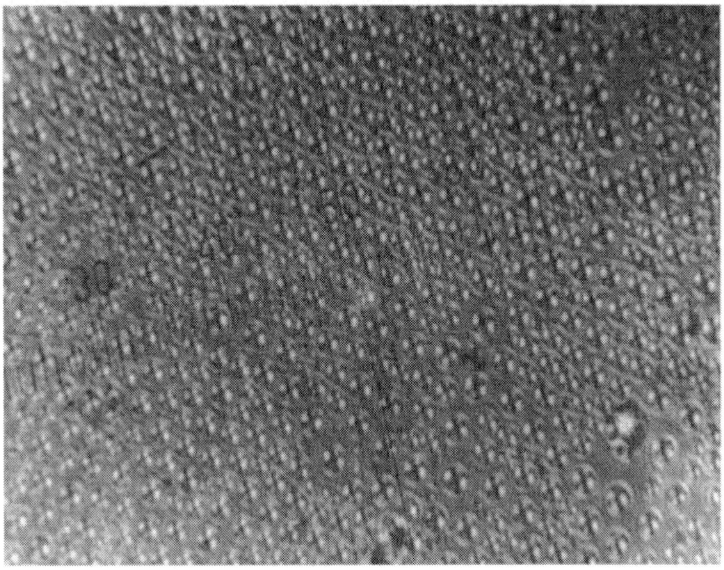

Figure 1.20a. Pre-transitional arrays of cross-like nematic domains and smectic-A undulations obtained in initially twisted nematic layer of 8CB with a thickness of 20 μm on cooling. A voltage of $10V_{r.m.s.}$ and a frequency of 10 kHz was applied across the nematic and the smectic-A phases, P⊥A, 10 small divisions correspond to 29 μm.

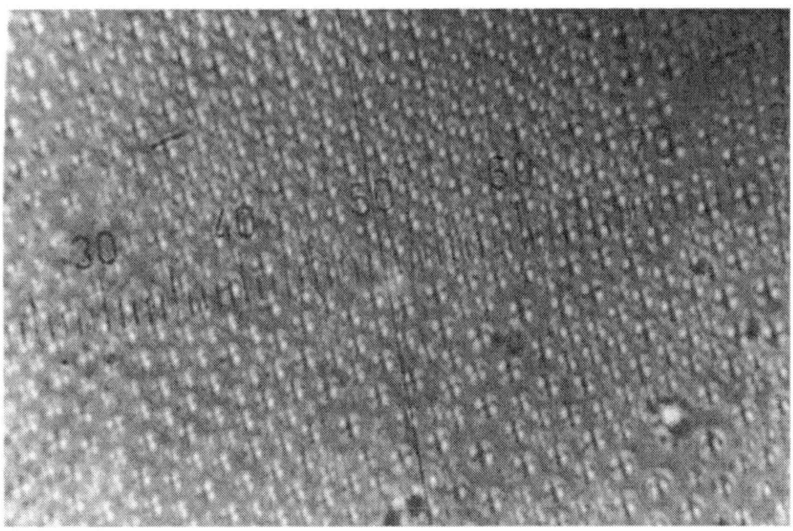

Figure 1.20b. The same as in Figure 1.20a after 15 s.

Examination of the photographs shows unambiguously that the cross-like domains are small nematic droplets, embedded into a slightly undulated smectic-A layer (see, for instance [313]). Further cooling of the texture led to the formation of complex smectic-A textures, shown in Figure 1.21a to Figure 1.21d.

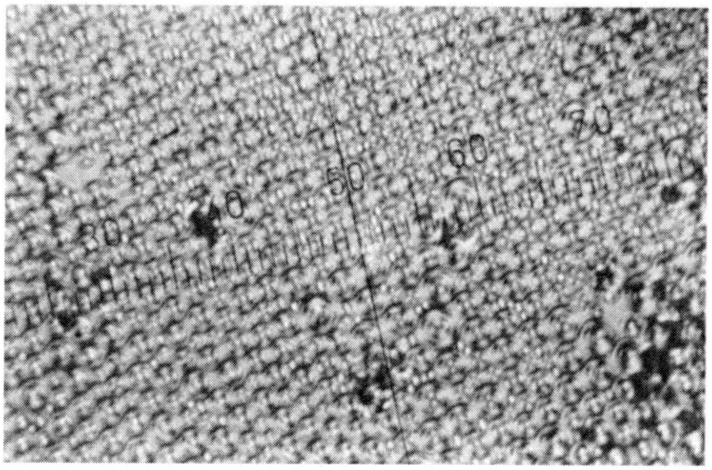

Figure 1.21a. Further cooling of the sample from Figure 1.20a, top focusing, the polarizer has been rotated along the clock's at an angle of 30°.

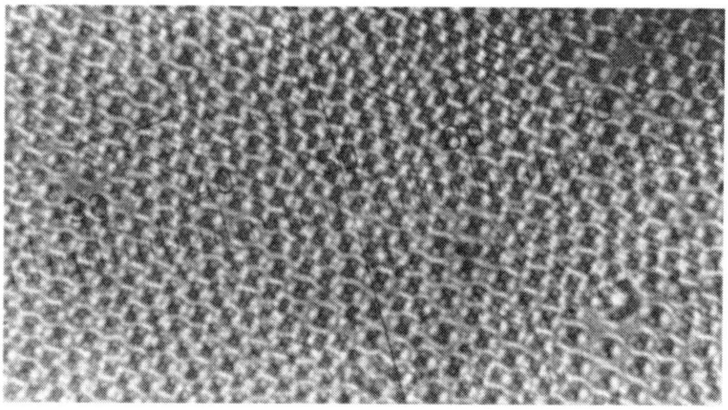

Figure 1.21b. The same as in Figure 1.21a, the polarizer has been rotated in the opposite direction at an angle of 30°.

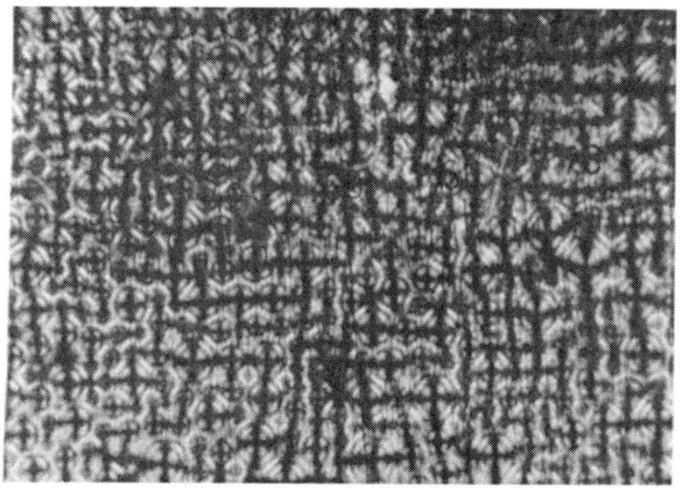

Figure 1.21c. The same as in Figure 1.21a. Down focusing of the light.

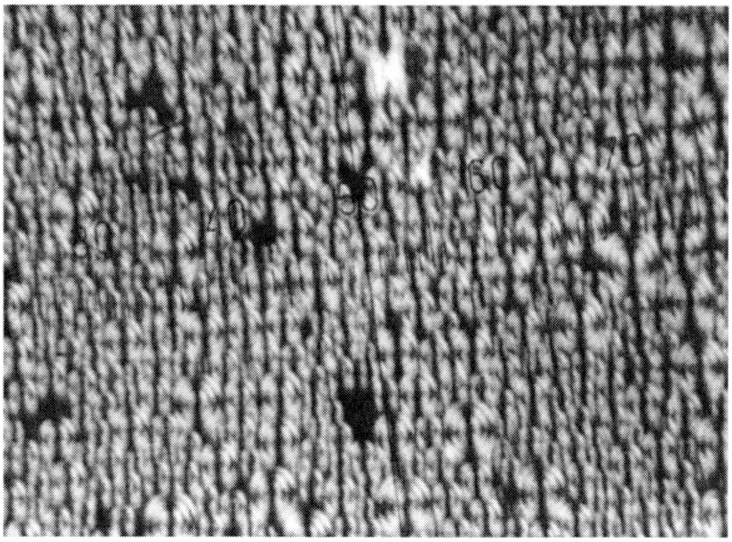

Figure 1.21d. The same as in Figure 1.21a. Middle focusing of the light.

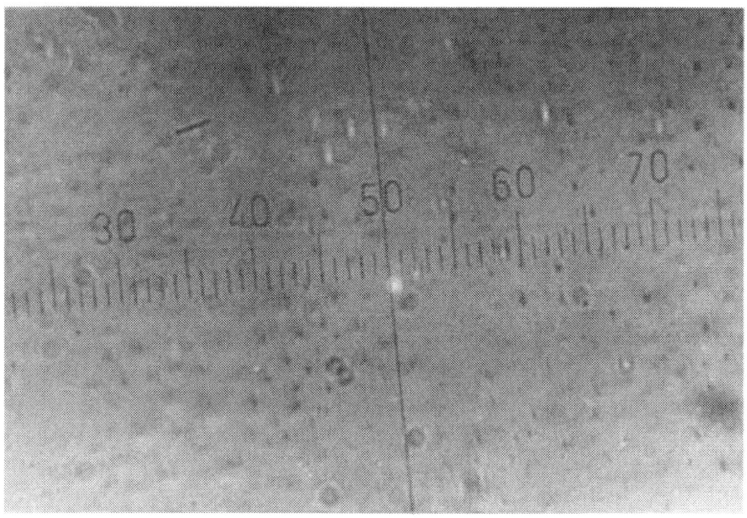

Figure 1.22. Disappearance of the confocal domains, embedded into a planar smectic-A layer (U=35$V_{r.m.s.}$, f=10 KHz); without an analyzer.

In our opinion, these images show that the smectic-A is deformed in the whole cell and the confocal domains with the smallest sizes are situated at both boundaries and the rest part of the smectic-A is undulated. Durand was the first author [314] who showed that undulation of the smectic-A layers at the boundary leads also to undulations of the other smectic-A layers in depth of the liquid crystal cell, which is not characteristic to the case of nematics and large-pitch cholesterics (for the undulations of the smectic-A layers see also References from [315] to [320]), It is clear that the undulations of the smectic-A layers are parallel to the surfaces of two glass plates confining the liquid crystal. Hartman [321] and Vertogen and de Jeu [322] proved that undulations of the smectic-A layers in a perpendicular direction would require a big elastic energy which should be distributed equally between the orientational (with respect to the smectic-A director) surface energy and the positional (with respect to the smectic-A layer) surface energy. Due to this large elastic energy, such possible smectic-A layer undulations are usually replaced by various kinds of confocal domains which have considerable smaller elastic energy. Textures, such as those shown in Figure 1.20 and Figure 1.21, have not been observed in the smectic-A phase of the liquid crystal 8CB, although this material has been widely studied. The smectic-A textures being formed can be visibly removed by application of an AC voltage with a sufficiently big amplitude between 20 $V_{r.m.s.}$ and 40 $V_{r.m.s.}$ (Figure 1.22).

At the decrease of the voltage, however, or at the switch-off of the voltage, the smectic-A textures reappear in the same kind. The electro-optic transient curves are shown in Figure 1.23.

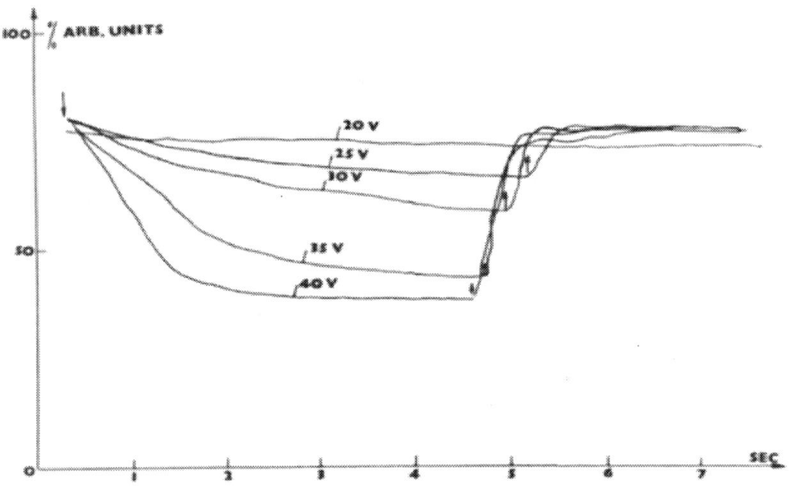

Figure 1.23. Electrooptic transmission curves of the light obtained with an electronic recorder for a smectic-A layer of 8CB with a thickness of 20 μm, crossed nicols.

The shape of these curves shows first, the electro-optic effect start with a threshold and second, that the electro-optic effect is completely reversible only in a narrow temperature interval of 3 to 5^0 C below the smectic-A – nematic phase transition, depending on the strength of the voltage, where the elastic behavior of the smectic-A still is manifested. Below this temperature range, the plastic character of the smectic-A prevails and the smectic-A textures are stored. Our experimental results clearly show that when the confocals are embedded into straight smectic-A layers, the electro-optic effects were not reversible and vice versa, when they are embedded into slightly undulated smectic-A layers the electro-optic effects are reversible. Further, in the dynamics of the electrooptic effects, the dislocations play an important role reducing the strains [168, 323]. According to Pershan [219], the position of the smectic-A defects with respect to the boundaries is important for their interaction with them. For instance, when the smectic-A defects are parallel to the boundaries, the smectic-A can be regarded as a free layer. In addition, the nucleation of the smectic-A phase from one deformed nematic should be, as noted above, heterogeneous [171, 192] as later proved by Kléman and Lavrentovich [195]. Furthermore, the value of the smectic constant of contraction or dilation of the smectic-A layers, depends considerably on the density of the smectic-A defects [191, 192, 219]. Lower value of the contraction (dilation) constant will facilitate the realization of one reversible electro-optic effect in the smectic-A. Bartolino and Durand [222] proved that the relaxation goes in two steps: first (for parts of milliseconds) and slow, with a time duration of several seconds. It is clear that these characteristics are not convenient for application in displays due to the slow relaxations and the narrow working smectic-A interval which is near the smectic-A – nematic phase transition. However, they can aid the study of the dynamics of confocal domains and dislocations. It is possible to obtain a reversible electrooptic effect in the case when the confocal domains are embedded into waved smectic-A layers, as noted above, since it is possible to preserve the linear dislocations in these structures at the application of a high-frequency voltage of a sufficient large amplitude whereas the rest part of the confocal domains and the wavy structures disappear. This experimental finding permits the confocal domains to return to their initial position at removal of the voltage. Moreover, in our opinion, the dislocations in the smectic-A phase play an important role in the dynamics of the electro-optical effects. According to our experimental results, the movement of the dislocations upon the application of a voltage is from the waved part of the liquid crystal to the centers of the focal-conics. The movement of the dislocations is in the opposite direction at the removal of the voltage. Such a behavior of the smectic-A texture is possible when the smectic-A domains are connected to surface irregularities on the glass plates or on the orienting SiO layer. The second important comment concerns the dynamic characteristics of the new electro-optic effect under consideration. It is clear from the electro-optic curves that the rise time when the voltage is switched on is much bigger, in the range of seconds. In the opinion of Ribotta and Durand [207], the relaxation of the local stresses in the smectic-A by the appearance and disappearance of dislocations are in the range of 50 ms, whereas the transit time of the dislocations coming from the bulk of the liquid crystal is in the range of seconds. Our experimental results show that the relaxation of the smectic-A textures is approximately milliseconds at switch-off of the voltage. It is clear that this time is connected by movement of some of the dislocations, which travel a very short distance in the range of a micron.

Consequently, the fast appearance of the smectic-A textures is governed by the surfaces whereas the disappearance of these smectic-A textures is governed by the bulk behavior. Third, the surface φ-elastic energy is considerably smaller than the bulk elastic energy of the focal conics and consequently is not important for the realization of reversible electro-optic effect in the smectic-A. Inversely, the θ-surface energy of the smectic-A is very important and it must be larger as compared to the total elastic energy of confocals. In this case, the confocals can be "caught" by the boundaries of the liquid crystal cells. From the literature we know that the various dislocations or focal-conic domains can easily glide between the smectic-A layers. According to Friedel [171], the regions of the focal conics when they move, have a solid friction which is of the order of the elastic energy stored in the production of the domains. This process is possible due to the surface anchoring of the liquid crystal.

The formation of the smectic-A phase, from the nematic twisted by $\pi/2$ with the aid of rubbed thin PVA layers and subjected under appropriate AC voltage under cooling is quite different. The liquid crystal was placed between two thin oxide-coated glass plates separated by Teflon spacers with a thickness of 10 μm or 20 μm.

The glass plates were dipped into a one weight percent aqueous solution of polyvinyl alcohol (PVA), removed slowly and dried in an oven at a temperature between 120 and 140^0 C. After the formation of the PVA layers, they were rubbed to produce grooves which determine the uniform homogeneous alignment of the liquid crystal. The liquid crystal had a high purity, the specific conductivity of the cells was measured to be around 2-3x10^{-12} $(\Omega cm)^{-1}$. Consequently, the rubbing of the PVA layers does not contaminate the liquid crystal. The cell was constructed in such a way as to ensure a twist orientation of the nematic. The good twist orientation of the nematic was observed under a polarizing microscope with a parallel position of the two nicols.

The nematic layer was subjected to an AC voltage with a frequency between 1 and 10 kHz and an amplitude between $30U_c$ and $25U_c$, where U_c is the threshold voltage which causes the appearance of the Fréedericksz transition. The threshold voltage in our case was measured to be 1.4 $V_{r.m.s.}$ which is slightly lower relative to that for the treatment of the glass plates by SiO (1.6 $V_{r.m.s.}$). The textures were quite different and are illustrated on Figure 1.24 and can be controlled electrically only close to the smectic-A – nematic phase transition, in the smectic-A phase (the time difference between two successive curves is 0.5 s).

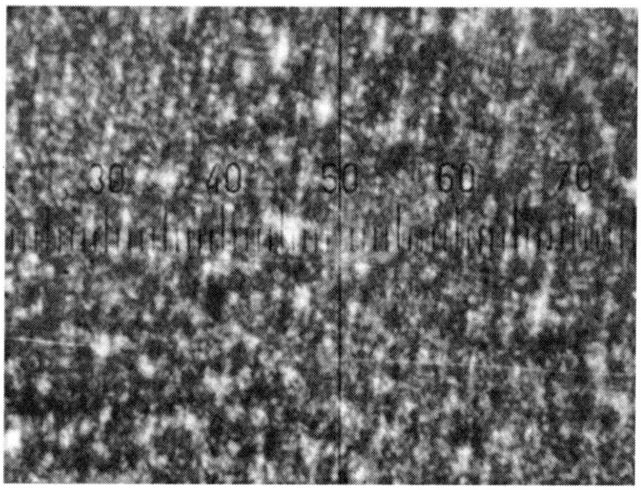

Figure 1.24. Strong-scattering smectic-A textures obtained in a 20 µm thick smectic-A layer of 8CBat a removing of the voltage of $40V_{r.m.s.}$, f=10 kHz, crossed nicols. The glass plates have been treated by PVA layer.

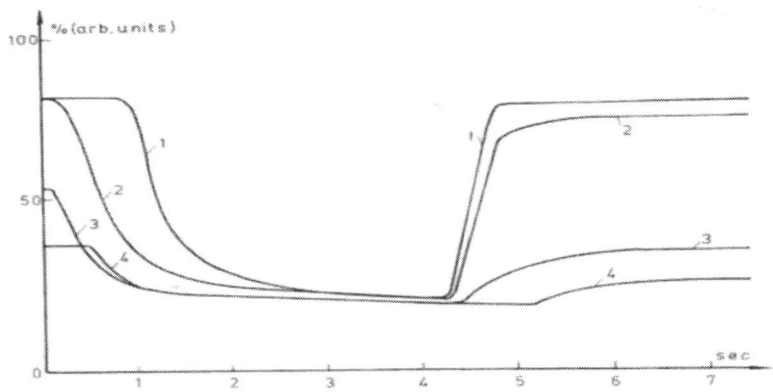

Figure 1.25. A reversible electro-optic behavior of 8BC smectic-A layer at the smectic-A- nematic phase transition (d=10 µm, U=$32V_{r.m.s.}$). The four electro-optic transmission curves have been obtained at a free cooling in the range of 30 s of the smectic-A layer, crossed nicols.

Let us now compare the electro-optic effect obtained earlier for a preceding nematic phase twisted by SiO with that obtained by the PVA anchoring. First, there is a large difference in the type of the smectic-A textures. With the SiO anchoring, the smectic-A textures are very regular with the presence of many focal conic domains which are embedded into a deformed smectic-A phase. With PVA anchoring, the smectic-A textures are very irregular with the presence of many defects which evidently lead to the observation of many irregular bright spots. Second, there is a large difference in the manner of the appearance or disappearance of the smectic-A textures. With SiO anchoring, the smectic-A textures appear or disappear very smoothly without any visible electrohydrodynamic movement. With PVA anchoring, the appearance or disappearance of the smectic-A textures is accompanied by a visible movement of the fluid in the planes of the Electrodes. As mentioned, the movement is caused by the smectic-A defects [189, 268, 270, 324] which are not pinned on the glass plates [191]. Evidently, the SiO anchoring layer has many irregularities which on the one hand can pin the smectic-A defects [191] and on the other hand aids for the heterogeneous nucleation of the focal conics [219]. The rubbed PVA layer is, however, smooth and the greater part of the dislocations and the disclinations are parallel to the glass plates since for such a position of the smectic-A defects, the value of the surface energy in the smectic-A phase is not important [219] (it is much less than 400 erg/cm^2), the smectic-A phase is free and the electro-hydrodynamic movement of the dislocations is in the plane of the Electrodes. We stress here that after the appearance or disappearance of the smectic-A scattering centers, the movement stops. The electro-hydrodynamic movement of the smectic-A due to the conductivity mechanism is usually local and continuous in time. For SiO anchoring, the greater part of the dislocations and disclinations are pinned on the glass plates and the anchoring of the smectic can influence only the deformation or orientation of the smectic-A. Finally, the difference in the two electro-optic effects might be due to the change of the elastic properties of the smectic-A in the presence of many defects. This effect, called “diaelasticity” has been suggested by Pershan [291] and studied by Pershan and Prost [191]. Further, the observed Grandjean walls for the PVA anchoring only (see Figure 1.4) show that the φ-azimuthal anchoring is stronger in comparison with SiO anchoring. Inversely, the polar θ-surface anchoring for the case of SiO anchoring is bigger relative to that in the case of PVA anchoring. Consequently, confocal domains for the case of PVA anchoring were observed only in the bulk and they easily have been replaced by the strongly-scattering smectic-A textures. Furthermore, at weak θ-anchoring, one observes a great tilt of the director which does not permit also the formation of circular-line focal conics.

In the end, we will comment briefly the possibilities for the application of the present effect. Evidently, it can be used in liquid crystal devices with storage. Furthermore, the smectic-A scattering textures observed in our experiments are very stable with respect to the temperature, even near to the smectic-A –nematic phase transition. On the other hand, the reversibility of the effect can be widened considerably by applying a voltage which is sufficiently high (for instance, in the range of 150-200 $V_{r.m.s.}$) which, however, should not be greater than that causing the complete orientation of the nematic, respectively smectic-A phases, or by decrease of the thickness of the smectic-A cells. Furthermore, various levels of contrast can be achieved by several applications of one and the same voltage. This is possible because of the partly-reversible character of this electro-optic effect in a certain temperature interval.

CONCLUSION

The nematic-smectic-A phase transition has been studied for many liquid crystals and one part of them shows a phase transition of a second order, or a very weak first order. In the basic theory of this phase transition, suggested by de Gennes, the elastic energy in analogy with the normal metal – superconductor phase transition, consists of two parts: the elastic energy of Frank and the elastic energy of the smectic-A, respectively with the aid of the complex order parameter Ψ. De Gennes showed that the connection between the fluctuations of the nematic director, usually designated by the unit vector **n**, and the complex smectic-A order parameter Ψ, leads to an increase of the value of the elastic constants of bend K_{33} and twist K_{22} which diverge at the approach of the nematic-smectic-A phase transition in the following manner:

$$K_{33} = K_{33}{}^{\circ} + \left(ckT\varsigma_{II} / d^2\right)$$

$$K_{22} = K_{22}{}^{\circ} + \left(ckT\varsigma_{\perp}{}^2 / d^2\varsigma_{II}\right)$$

where $\zeta_{||}$is the longitudinal smectic-A correlation length, parallel to the nematic director and $\zeta_{\perp}$ is the transversal smectic-A corre;ation length, normal to the nematic director, kT is the thermal energy, c is a constant of the order of unity and d is the distance between the smectic-A layers.

On the other hand, the rotational viscosity γ_1 diverges near the nematic-smectic-A transition according to the following formula:

$$\gamma_1 = \gamma_1^{\circ} + c'\gamma_3 kT\varsigma / c_{11}$$

In a number of papers and certificates [66-70, 325, 326] we suggested a new method for the growth of the smectic-A phase from a strongly-deformed in the boundary regions nematic layer which can be splay-bended, splay-twisted, etc. This can be achieved by the application of a voltage with a sufficiently high frequency in order to avoid the appearance of the electro-hydrodynamics and with a sufficiently high amplitude in order to bring about strong deformations of the nematic in the boundary regions followed by cooling of the nematic, in the presence of the voltage, to the smectic-A phase. Then the voltage is removed after the formation of the various smectic-A textures. The novelty of this method is that the nematic-smectic-A phase transitional temperature in the boundary regions decreases with $1\text{-}2^0$ C due to the strong deformations in the boundary regions of the nematic. At a homogeneous distribution of the temperature along Z (i.e., normal to the glass plates), the smectic-A phase grows first in the middle part of the cell where the nematic is completely aligned and after that in the boundary regions. Such a decrease of the nematic-smectic-A phase transitional temperature was discovered theoretically by de Gennes [13]. This manner of the smectic-A growing can give information about the matching of the smectic-A in the boundary regions with the middle oriented smectic-A only for the case of smectics-A which are of the first kind when $K < 1/\sqrt{2}$. Chandrasekhar [16] determined $K = \lambda/\xi$ as a parameter of Ginzburg-Landau, where λ is the depth of penetration of the deformations of bend or splay and ξ is the coherence length of deformations penetration in the smectic phase. It is clear that opposite inequality K>1/√2 determines smectics of a second order which are not studied here. The simultaneous existence of the nematic and the smectic-A phases around the nematic-smectic-A phase transition of a second or a weakly first order with enthalpy of the transition smaller than kT continue to provoke interest and to stimulate towards new theoretical and experimental studies [327, 328]. The formation of various confocal structures in the boundary regions permits to estimate also the smectic-A surface energy which is not hitherto sufficiently studied. The very arrangement of the confocal domains can be used in diffraction images and even in photonic structures [329, 330].

REFERENCES

[1] Kelker, H. *Mol. Cryst. Liq. Cryst.* 1988, 165, 1-43.
[2] Reinitzer, F. *Monatshefte für Chemie* 1888, 9, 421-441, "Contributions to the knowlidge for cholesterol", English Translation in *Liq. Cryst.(UK)* 1989, 5, 7-18.
[3] Lehman, O. *Z. Phys. Chem.* 1889, 4, 462-472.
[4] Sackman, H.; Demus, D. *Mol. Cryst. Liq. Cryst.* 1973, 21, 239-273.
[5] Demus, D. *Mol. Cryst. Liq. Cryst.* 1988, 165, 45-84.
[6] Chandrasekhar, S. *Contemp. Phys.* 1988, 29, 527-558.
[7] Demus, D. *Mol. Cryst. Liq. Cryst.* 1989, 5, 75-110.
[8] Coates, D.; Gray, G. W. *Microscope* 1971, 24, 117-150.
[9] Demus, D.; Diele, S.; Grande, S.; Sackman, H, In *Advances in Liquid crystals*; Brown, G. H.; Ed.; Academic Press: Orlando, US, 1983, 1-107.
[10] Durand, G.; Litster, J. D. *Ann. Rev. Mat. Sci.* 1973, 3, 269-292.
[11] Litster, J. D. *Phil. Tran. R. Soc. Lond. A* 1983, 309, 145-153.
[12] De Vries, A.; Ekachai, A.; Spielberg, N. *Mol. Cryst. Liq. Cryst. Lett.* 1979, 49, 143-152.
[13] De Gennes, P. G. *The Physics of Liquid crystals*; Clarendom Press: Oxford, England, 1974, chapter 4.
[14] Andereck, B. S. *Int. J. Mod. Phys. B* 1995, 9, 2139-2155.
[15] Anisimov, M. A.; Cladis, P. E.; Gorodetskii, E. E.; Huse, D. A.; Podneks, V. E.; Taratuta, V. G.; Van Saarloos, W.; Voronov, V. P. *Phys. Rev. A* 1990, 41, 6749-6762.
[16] Chandrasekhar, S. *J. Stat. Phys.* 1984, 34, 883-893.
[17] Stephen, M. J.; Straley, J. P. *Rev. Mod. Phys.* 1974, 46, 617-704.
[18] Guyon, E. *Am. J. Phys.* 1975, 43, 877-881.
[19] Litster, J. D.; Birgeneau, R. *Physics Today* 1982, N 5, 1-8.
[20] Litster, J. D.; Garland, C. W.; Lushington, K. J.; Schaetzing, R. *Mol. Cryst. Liq. Cryst.* 1981, 63, 145-156.
[21] Lubensky, T. C. *J. Chim. Phys.* 1983, 80, 31-43.
[22] McMillan, W. L. *J. Phys. Colloq. (Paris)* 1975, 36, C1-103 – C1-105.
[23] Prost, J. *Adv. Phys.* 1984, 33, 1-46.
[24] Thoen, J. *Int. J. Mod. Phys. B* 1995, 9, 2157-2218.
[25] Kobayashi, K. K. *J. Phys. Soc. Jpn.* 1970, 29, 101-105.
[26] McMillan, W. L. *Phys. Rev. A* 1971, 4, 1239-1246.

[27] Chandrasekhar, S. *Liquid crystals* 1sted. Cambridge University Press, UK, 1977, chapter 5.
[28] Maier, W.; Saupe, A. *Z. Naturforsch. (a)* 1960, 15, 287-292.
[29] Le Grange, J. D.; Mochel, J. M. *Phys. Rev. A* 1981, 23, 3215-3223.
[30] De Gennes, P. G. *Sol. St. Commun.* 1972, 10, 753-756.
[31] Halperin, B. I.; Lubensky, T. C. *Sol. St. Commun.* 1974, 14, 997-1001.
[32] Halperin, B. I.; Lubensky, T. C.; Ma, Sh.-k, *Phys. Rev. Lett.* 1974, 32, 292-295.
[33] Shiwa, Y. *Prog. Theor. Phys.* 1976, 55, 629-630.
[34] Albertini, G.; Corinaldesi, M.; Mazkedianm F.; Melone, S.; Ponzi-Bossi, M. G.; Rustchelli, F. *Sol. St. Commun.* 1977, 24, 433-437.
[35] Brochard, F. *J. Phys. Colloq. (Paris)* 1973, 34, 411-422.
[36] Vithana, H. K. M.; Surendranath, V.; Lewis, M.; Baldwin, A.; Eidner, K.; Mahmood, R.; Johnson, D. L. *Phys. Rev. A* 1990, 41, 2033-2037.
[37] Yethiraj, A,; Bechhoefer, J. *Mol. Cryst. Liq. Cryst.* 1997, 304, 301-307.
[38] Vithana, H. K. M.; Xu, G.; Johnson, D. L. *Phys. Rev. E* 1993, 47, 3441-3455.
[39] Jähnig, F.; Brochard, F. *J. Phys. (Paris)*1974, 35.301-313.
[40] Léger, L.; Martinet, A. *J. Phys. Colloq. (Paris)* 1976, 37, C3-89 – C3-97.
[41] Rościszewski, K. *Acta Phys. Polonica A* 1981, 59, 611-618.
[42] Andereck, B. S.; Patton, B. R. *Phys. Rev. E* 1994, 49, 1393-1403.
[43] Brochard, F. *J. Phys. Colloq. (Paris)* 1976, 37, C3-85 – C3-88.
[44] Chen, J.-H.; Lubensky, T. C. *Phys. Rev. A* 1976, 14, 1202-1207.
[45] Chu, K. C.; McMillan, W. L. *Phys. Rev. A* 1977, 15, 1181-1187.
[46] Herbut, I. F.; Yethiraj, A.; Bechhoefer, J. *Europhys. Lett.* 2001, 55, 317-323.
[47] Lelidis, I. *Phys. Rev. Lett.* 2001, 86, 1267-1270.
[48] Lubensky, T. C.; Chen, J.-H. *Phys. Rev. B* 1978, 17, 366-376.
[49] Patton, B. R.; Andereck, B. S. *Phys. Rev. Lett.* 1992, 69, 1556-1559.
[50] Yethiraj, A., ; Bechhoefer, J. *Phys. Rev. Lett.* 2000, 84, 3642-3645.
[51] Madhusudana, N. V.; Srikanta, B. S. *Mol. Cryst. Liq. Cryst.* 1983, 99, 375-383.
[52] Cladis, P. E.; Torza, S. *J. Appl. Phys.* 1975, 46, 584-599.
[53] Lubensky, T. C. *J. Phys. Colloq. (Paris)* 1975, 36, C1-151 – C1-152.
[54] Crandall, K. A.; Rosenblatt, Ch.; Hornreich, R. M. *Liq. Cryst. (UK)* 1995, 18, 251-256.
[55] Ranganath, G. S. *Mol. Cryst. Liq. Cryst. Lett.* 1983, 92, 201-207.
[56] Durand, G. *Liq. Cryst. (UK)* 1993, 14, 159-168.
[57] Tamblyn, N.; Oswald, P.; Miele, A.; Bechhoefer, J. *Phys. Rev. E* 1995, 51, 2223-2236.
[58] Chu, K. C.; McMillan, W. L. *Phys. Rev. A* 1977, 15, 337-340.
[59] Majoros, S. J.; Brisbin, D. J.; Johnson, D. L.; Neubert, M. E. *Phys. Rev. A* 1979, 20, 1619-1627.
[60] Gooden, C.; Mahmood, R.; Brisbin, D.; Baldwin, A.; Johnson, D. L.; Neubert, M. E. *Phys. Rev. Lett.* 1985, 54, 1035-1038.
[61] Allender, D. W.; Hornreich, R. M.; Johnson, D. L. *Phys. Rev. Lett.* 1987, 59, 2654-2657.
[62] Chevalard, C.; Clerc, M. G. *Phys. Rev. E* 2001,65, 011708 (8pp).
[63] Wen, B.; Rosenblatt, Ch. *Phys. Rev. Lett.* 2002, 89, 195505 (4 pp).
[64] Demus, D.; Richter. L. *Textures of Liquid crystals*, VEB Deutscher Verlag für Grundstofindustrie, Leipzig, Germany; 1978, pp 204-205.

[65] Pergamenshchik, V. M.; Lelidis, I.; Uzunuva, V. A. *Phys. Rev. E* 2008, 77, 041703 (8 pp).

[66] Hinov, H. P. *J. Phys. (Paris)* 1981, 42, 307-315.

[67] Hinov, H. P. *Japan Display'83,* P215, 360-363.

[68] Hinov, H. P.; Shonova, N.; Avramova, K. *Mol. Cryst. Liq. Cryst.* 1983, 97. 297-315.

[69] Hinov, H. P. *Liq. Cryst. (UK)* 1988, 3, 1481-1503.

[70] Hinov, H. P.; Avramova, K. *Liq. Cryst. (UK)* 1988, 3, 1505-1514.

[71] Navard, P.; Haudin, J. M. *J. Thermal Analysis* 1985, 30, 61-64.

[72] Ratna, B. R.; Chandrasekhar, S. *Mol. Cryst. Liq. Cryst.* 1988, 162B, 157-159.

[73] Djurek, D.; Baturić-Rubčič, J.; Fraunulovic, K. *Phys. Rev. Lett.* 1974, 33, 1126-1129.

[74] Hardouin, F.; Gasparoux, H.; Delhaes, P. *J. Phys. Colloq. (Paris)* 1975, 36, C1-127 – C1-131.

[75] Bata, L.; Buka, A.; Jánossy, I.; Szabon, J. In *Advances in liquid crystal Research and Applivation*, Bata, L.; Ed., Pergamon Press: Oxford-Akademia Kiado, Budapest, Hungary, 1980, pp 209-219.

[76] Coles, H. J.; Strazielle, C. *Mol. Cryst. Liq. Cryst.* 1979, 55, 237-250.

[77] Kasting, G. B.; Garland, C. W.; Lushington, K. J. *J. Phys. (Paris)* 1980, 41, 879-884.

[78] Navard, P.; Cox, R. *Mol. Cryst. Liq. Cryst. Lett.* 1984, 102, 261-264.

[79] Cheung, L.; Meyer, R. B.; Grüler, H. *Phys. Rev. Lett.* 1973, 31, 349-351.

[80] Delaye, M.; Ribotta, R.; Durand, G. *Phys. Rev. Lett.* 1973, 31, 443-445.

[81] Cladis, P. E. *Phys. Rev. Lett.* 1973, 31, 1200-1203.

[82] Salin, D.; Smith. I. W.; Durand, G. *J. Phys. Lett. (Paris)* 1974, 35, L-165 – L-168.

[83] Cladis, P. E. *Phys. Lett. A* 1974, 48, 179-180.

[84] Madhusudana, N. V.; Karat, P. P.: Chandrasekhar, S. *Pramana, suppl. N 1* 1974, 225-236.

[85] Torza, S.; Cladis, P. *Phys. Rev. Lett.* 1974, 32, 1406-1409.

[86] Bacri, J. C. *J. Phys. Colloq. (Paris)* 1975, 36, C1-123 – C1-125.

[87] Hardouin, F.; Achard, M. F.; Gasparoux, H. *Sol. St. Commun.* 1974, 14, 453-456.

[88] Hardouin, F.; Achard, M. F.; Sigaud, G.; Gasparoux, H. *Phys. Lett. A* 1974, 49, 25-26.

[89] Huang, Cheng-Cher; Pindak, R. S.; Flanders, Ph. J.; Ho, J. T. *Phys. Rev. Lett.* 1974, 33, 400-402.

[90] Chu, K. S.; McMillan, W. L. *Phys. Rev. A* 1975, 11, 1059-1067.

[91] Wise, R. A.; Olah, A.; Doane, J. W. *J. Phys. Colloq. (Paris)* 1975, 36, C1-117 – C1-120.

[92] Gasparoux, H.; Hardouin, F.; Achard, M. F.; Sigaud, G. *J. Phys. Colloq. (Paris)* 1975, 36, C1-107 – C1-111.

[93] Langevin, D. *J. Phys. (Paris)* 1976, 37, 901-907.

[94] Kiry, F.; Martinoty, P. *J. Phys. Colloq. (Paris)* 1976, 37, C3-113 – C3-117.

[95] Sakamoto, A.; Yoshino, K.; Kubo, U.; Inuishi, Y. *Jpn. J. Appl. Phys.* 1976, 15, 745-746.

[96] Als-Nielsen, J.; Birgeneau, R. J.; Kaplan, M.; Litster, J. D.; Safinya, C. R. *Phys. Rev. Lett.* 1977, 39, 352-355.

[97] Birecki, H.; Litster, J. D. *Mol. Cryst. Liq. Cryst.* 1977, 42, 33-41.

[98] Martins, A. F.; Diogo, A. C.; Vaz, N. P. *Ann. Phys. (Fr.)* 1978, 3, 361-368.

[99] Bahadur, B. *Phys. Lett. A* 1975, 55, 133-134.

[100] Demus, D.; Deutscher, H.-J.; König, S.; Kresse, H.; Kuschel, F.; Pelzl, G.; Schubert, H.; Selbmann, Ch.; Weissflog, W.; Wiegeleben, A.;, Wulf, J. In *Forschungen über Flüssige Kristalle,* Demus, Ed.; Martin-Luther University Halle (Germany): *Wiss. Beitrage* 1978, 21, 9-15.
[101] Sarna, R. K.; Bhide, V. G.; Bahadur, B. *Mol. Cryst. Liq. Cryst.* 1982, 88, 65-79.
[102] Léger, L.; Martinet, A. *J. Phys. Colloq. (Paris)* 1976, 37, C3-89 – C3-97.
[103] Leadbetter, A. J.; Durrant, J. L. A.; Rugman, M. *Mol. Cryst. Liq. Cryst. Lett.* 1977, 34, 231-235.
[104] Jánossy, I.; Bata, L. *Acta Phys. Polonica* 1978, 54, 643-648.
[105] Davidov, D.; Safinya, C. R.; Kaplan, S. S.; Dana, S. S.; Schaetzing, R.; Birgeneau, R. J.; Litster, J. D. *Phys. Rev. B* 1979, 19, 1657-1663.
[106] Coles, H. J.; Sefton, M. S. *Mol. Cryst. Liq. Cryst. Lett.* 1987, 4(5), 123-130.
[107] Zakharov, A. V.; Thoen, J. *Phys. Rev. E* 2004, 69, 051709 (4pp).
[108] Litster, J. D.; Als-Nielsen, J.; Birgeneau, R. J.; Dana, S. S.; Davidov, D.; Garcia-Golding, F.; Kaplan, M.; Safinya, C.R.; Schaetzing, R. *J. Phys. Colloq. (Paris)* 1979, 40, C3-339 – C3-344.
[109] Lewis, M. E.; Khan, I.; Vithana, H.; Baldwin, A.; Johnson, D. L.; Neubert, M. E. *Phys. Rev. A* 1988, 38, 3702-3709.
[110] Escher, C.; Wingen, R. *Adv. Mater.* 1992, 4, 189-190.
[111] Mi, X.-D.; Yang, D.-K. *Phys. Rev. E* 1998, 58, 1992-2000.
[112] Janning, J. L. *Appl. Phys. Lett.* 1972, 21, 173-174.
[113] Guyon, E.; Pieranski, P.; Boix, M. *Lett. Appl. Eng. Sci.* 1973, 1, 19-24.
[114] Yamashita, M.; Amemiya, Y. *Jpn. J. Appl. Phys.* 1976, 15, 2087-2092.
[115] Pollack, J. M.; Haas, W. E.; Adams, J. E. *J. Appl. Phys.* 1977, 48, 831-833.
[116] Goodman, L. A.; McGinn, J. T.; Anderson, Ch. H.; Divergonimo, F. *IEEE Tr. El. Dev.* 1977, ED-24, 795-804.
[117] Aoyama, H.; Yamazaki, Y.; Matsuura, N.; Mada, H.; Kobayashi, Sh. *Mol. Cryst. Liq Cryst. Lett.* 1981, 72, 127-132.
[118] Kutty, T. R. N.; Fischer, A. G. *Mol. Cryst. Liq. Cryst.* 1983, 99, 301-318.
[119] Geary, J. M.; Goodby, J. W.; Kmetz, A. R.; Patel, S. *J. Appl. Phys.* 1987, 62, 4100-4108.
[120] Vetter, P.; Ohmura, Y.; Uchida, T. *Jpn. J. Appl. Phys.* 1993, 32, L1239-L1241.
[121] Faetti, S.; Nobili, M. *Phys. Lett. A* 1996, 217, 133-136.
[122] Kubono, A.; Onoda, H.,; Inoue, K.; Tanaka, K.; Akiyama, R. *Mol. Cryst. Liq. Cryst.* 2002, 373, 127-141.
[123] Uchida, T.; Ohgawara, M.; Shibata, Y. *Mol. Cryst. Liq. Cryst.* 1983, 98, 149-161.
[124] Sanda, P. N.; Dove, D. B.; Ong, H L.; Jansen, S. A.; Hoffmann, R. *Phys. Rev. A* 1989, 39, 2653-2658.
[125] Brochard, L. F., Orsay Liquid crystal Group *Contemp. Phys.* 1977, 18, 247-264.
[126] Hiltrop, K.; Stegemeyer, H. *Ber. Bunsenges. Phys. Chem.* 1978, 82, 884-889.
[127] Hiltrop, K., Stegemeyer, H. *Ber. Bunsenges. Phys. Chem.* 1981, 85, 582-588.
[128] Hiltrop, K.; Hasse, J.; Stegemeyer, H. *Ber. Bunsenges. Phys. Chem.* 1994, 98, 209-213.
[129] Gogolewski, S. *J. Phys. E: Sci. Instrum.* 1975, 8, 506-507.
[130] Neubert, M. E.; Norton, P.; Fishel, D. L. *Mol. Cryst. Liq. Cryst.* 1975, 31, 253-257.
[131] Yokoyama, H.; Kobayashi, Sh.; Kamei, H. *Rev. Sci. Instrum.* 1983, 54, 611-612.
[132] Adams, R. J. *J. Phys. E: Sci. Insrum.* 1987, 20, 420-423.

[133] Balzer, W.; Tschudi, T. *J. Phys. E: Sci. Instrum.* 1987, 20, 568-571.
[134] Born, M.; Wojff, E. *Principles of Optics*; Pergamon Press: Oxford, UK, 1968.
[135] Hartshorne, N. H.; Stuart, A. *Crystals and the Polarizing Microscope;*Elsevier: New York 1969.
[136] Chang, R. *Mat. Res. Bull.* 1972, 7, 267-278.
[137] Cloud, S. D. *Am. J. Phys.* 1973, 41, 1184-1188.
[138] Robert, J.; Ghardjedaghi, F. *C. R. Acad. Sc. Paris B* 1974, 278, 73-76.
[139] Hartshorne, N. H. In *Liquid crystals and Plastic Crystals*; Gray, G. W., Winsor, P. A.; Eds., Ellis Horwood Lmtd: Chichester, U.K., vol. 2, ch. 2, pp 24-61.
[140] Meyerhofer, D. *J. Appl. Phys.* 1977, 48, 1179-1185.
[141] Baum, S. J. *Opt. Eng.* 1977, 16, 291-294.
[142] Carr, E. F.; McClymer, J. P. *Am. J. Phys.* 1991, 59, 366-367.
[143] Lavrentovich, O. D. *Pramana J. Phys.* 2003, 61, 373-384.
[144] Huang, C. C.; Pindak, R. S.; Ho, J. T. *J. Phys. Lett. (Paris)* 1974, 35, L-185 – L-187.
[145] Kim, M. G.; Park, S.; Cooper, M; Letcher, S. V. *Mol. Cryst. Liq. Cryst.* 1978, 36, 143-152.
[146] Ratna, B. R.; Shashidhar, R. *Mol. Cryst, Liq. Cryst.* 1977, 42, 113-123.
[147] Belardi, Z,; Guilland, G.; Maitrot, M.; Huck, J.; Simon, J.; Tournilhac, F. *Rev. Phys. Appl.* 1988, 23, 143-147.
[148] Barnik, M. I.; Beljaev, S. B.; Rumjanzev, B. G.; Tsvetkov, B. A.; Shitkov, N. M. In *Forschungen über Flüssige Kristalle 1.* NPOB D. Demus; Ed., Marthin-Luther Univerity Halle-Witenberg, *Wiss. Beitrage* 1978, 21, 84-94.
[149] Ratna, B. R.; Shashidhar, R. *Pramana J. Physics* 1976, 6, 278-283.
[150] Dunmur, D. A.; Manterfield, M. R.; Miller, W. H.; Dunleavy, J. K. *Mol. Cryst. Liq. Cryst.* 1978, 45, 127-144.
[151] Horn, R. G. *J. Phys. (Paris)*1978, 39, 105-109.
[152] Madhusudana, N. V.; Pratibha, R. *Mol. Cryst. Liq. Cryst.* 1982, 89, 249-257.
[153] Thoen, J.; Menu, G. *Mol. Cryst. Liq. Cryst.* 1983, 97, 163-176.
[154] Hakemi, H.; Jagadzinski, E. F.; DuPre D, *J. Chem. Phys.* 1983, 78, 1513-1518.
[155] Fisch, M. R.; Pershan, P. S.; Sorensen, L. B. *Phys. Rev. A* 1984, 29, 2741-2750.
[156] Morris, S. W.; Palffy-Muhoray, P.; Balzarini, D. A. *Mol. Cryst. Liq. Cryst.* 1986, 139, 263-280.
[157] Chmielewski, A. G. *Mol. Cryst. Liq. Cryst.* 1986, 132, 339-352.
[158] Yun, C. K.; Picot, J. J. C.; Predrickson, A. G. *J. Appl. Phys.* 1971, 42, 4764-4767.
[159] Rajan, V. S. V.; Picot, J. J. C. *Mol. Cryst. Liq. Cryst.* 1972, 17, 109-119.
[160] Rajan, V. S. V.; Picot J. J. C. *Mol. Cryst. Liq. Cryst.* 1973, 20, 55-68.
[161] Urbach, W.; Hervet, H. Rondelez, F. *Mol. Cryst. Liq. Cryst.* 1978, 46, 209-221.
[162] Rondelez, F.; Urbach, W.; Hervet, H. *Phys. Rev. Lett.* 1978, 41, 1058-1062.
[163] Nagasaka, Y.; Nagashima, A. *J. Phys. E: Sci. Instrum.* 1981, 14, 1435-1440.
[164] Nagasuka, Y.; Nagashima, A. *Rev. Sci. Instrum.* 1981, 52, 229-232.
[165] Cladis, P. E.; Van Saarloos, W.; Huse, D. A.; Patel, J. S.; Goodby, J. V.; Finn, P. L. *Phys. Rev. Lett.* 1989, 62, 1764-1767.
[166] Hinov, H. P.; Pisipati, V. G. K. M. *Liq. Cryst. (UK)* 1990, 8, 193-209.
[167] De Vries, A. *Mol. Cryst. Liq. Cryst.* 1981, 63, 215-230.
[168] Bartolino, R.; Durand, G. *Nuovo Cimento D* 1984, 3, 903-913.
[169] Cagnon, M.; Durand, G. *Phys. Rev. Lett.* 1993, 70, 2742-2745.

[170] Durand, G. *J. Chim. Phys.* 1983, 80, 119-125.
[171] Friedel, J. *J, Phys. Colloq. (Paris)* 1979, 40, C3-45 – C3-52.
[172] Ishikawa, T,; Lavrentovich, O. D., In *Defects in Liquid crystals, Computer Simulation, Theory and Experiments,* Lavrentovich, O. D.; Ed., Kluwer Academic Publisher: Amsterdam, Neherland 2001, pp 271-301.
[173] Kléman, M.; Michel, L. *J. Phys. Lett.)Paris)* 1982, 39, L-29 – L-32.
[174] Oswald, P., Kléman, M. *J. Phys, Lett. (Paris)*1982, 43, L-411 – L-415.
[175] Wojtowicz, P. J. RCA *Rev.* 1974, 35, 384-406.
[176] Parodi, O. *Sol. St. Commun.* 1972, 11, 1503-1507.
[177] Goscianski, M.; Léger, L.; Mirsea-Roussel, A. J. Phys. Lett. (Paris) 1975, 36, L-313 – L- 316.
[178] Ktorides, P. I.; Uhrich, D. L. *Mol. Cryst. Liq. Cryst.* 1982, 87, 69-76.
[179] Kléman, M.; Lavrentovich, O. D. *Eur. Phys. J. E* 2000, 2, 47-57.
[180] Meunier, J. P.; Billard, J. *Mol. Cryst. Liq. Cryst.* 1969, 7, 421-431.
[181] Friedel, J.; Kléman, M. In *Fundamental Aspects of Dislocations*; Simons, J. A.; De Wit, R.; Bullongh, R.; Eds.; Nat, Bur, Stand.: U.S., Spec. Publ. 317, I 1970, pp 607-636.
[182] Geurst, J. A. *Phys. Lett. A* 1971, 34, 283-284.
[183] Bouligand Y. *J. Phys.(Paris)* 1972, 33, 525-547.
[184] Rault, J. *Phil. Mag.* 1976, 34, 753-765.
[185] Kléman, M. *Phil. Mag.* 1976, 34, 79-87.
[186] Trebin, H.-R.; Kutka, R. *J. Phys. Lett. (Paris)* 1981, 42, L-421 – L-423.
[187] Lavrentovich, O. D.; Kléman, M.; Pergamenshchik, V. M. *J. Phys, II France* 1994, 4, 377-404.
[188] Kléman, M.; Lavrentovich, O. D. *Liq. Cryst. (UK)* 2009, 36, 1085-1099.
[189] Le Berre, S.; Hareng, M. *Etudes desomene basse frequence dans les smectiques A*, Thomson-CSF Lab. Central de Rechershes 1977, pp 1-48.
[190] Taylor, G. N.; Kahn, F. J. *J. Appl. Phys.* 1974, 45, 4330-4338.
[191] Pershan, P. S.; Prost, J. *J. Appl. Phys.* 1975, 46, 2343-2353.
[192] Chou, N. J.; Depp, S. W.; Eldrige, J. M.; Lee, M. H.; Sprokel, G.; Juliana, A., Brown, J. *J. Appl. Phys.* 1983, 54, 1827-1832.
[193] Jähnig, F. *J. Phys. (Paris)* 1975, 36, 315-324.
[194] Huang, C. C.; Pindak, R. S.; Ho, J. T. *Sol. St. Commun.* 1978, 25, 1015-1917.
[195] Lavrentovich, O. D.; Kléman, M. *Phys. Rev. E* 1993, 48, R39-R42.
[196] Bidaux, R.; Boccara, N.; Sarma, G.; De Seze, L.; De Gennes, P. G.; Parudi, O. *J. Phys. (Paris)* 1973, 34, 661-672.
[197] Sethna, J. P.; Kléman, M. *Phys. Rev. A* 1982, 26, 3037-3040.
[198] Lavrentovich, G. D. *Mol. Cryst. Liq. Cryst.* 1987, 151, 417-424.
[199] Aste, T. *Phys, Rev. E* 1996, 53, 2571-2579,
[200] Blanc, Ch.; Kléman, M. *Phys. Rev. E* 2000, 62, 6739=6748.
[201] Baram, R. M.; Herrmann, H. J.; Rivier, N. *Phys. Rev. Lett.* 2004, 92, 044301 (4 pp).
[202] Perez, E.; Proust, E. *J. Phys. Lett. (Paris)* 1977, 38, L-117 - L -120.
[203] Niessen, A. K.; Den Ouden, A. *Philips Res. Reports* 1974, 29, 119-138,
[204] Boschmans, M. *Mol. Cryst. Liq. Cryst.* 1995, 261, 15-28.
[205] Galerne, Y. *J. Phys. Lett. (Paris)* 1979, 40, L-73 – L-76.
[206] Proust, J. E.; Perez, E. *J. Phys. Lett. (Paris)* 1977, 38, L-91 – L-94.
[207] Ribotta, R.; Durand, G. *J. Phys. (Paris)* 1977, 38, 179=204.

[208] Marignan, J.; Malet, G.; Parodi, O. *Ann. Phys. (France(* 1978, 3, 221-228.
[209] Marignan, J.; Parodi, O. *J. Phys. Colloq. (Paris)* 1979, 40, C3-78 – C3-82.
[210] Marignan, J.; Parodi, O. *J. Phys (Paris)*1983, 44, 665-673.
[211] Grandjean, F. *Bull. Soc. Franc. Min.* 1916, 39, 164-213.
[212] Friedel, G. *Ann. Phys. (Leipzig)* 1922, 18, 273-474.
[213] Kléman, M.; Williams, C. E. *Phil. Mag.* 1977, 35, 33-56.
[214] Benton, W. J.; Miller, C. A. *Prog. Coll. Polym. Sci.* 1983, 68, 71-81.
[215] De Gennes, P. G.*C, R. Acad. Sc. Paris B* 1972, 275, 939-941.
[216] Williams, C. E.; Kléman, M. *J. Phys. Lett. (Paris)* 1974, 35, L-33 – L-35.
[217] Kléman, M. *J. Phys. (Paris)* 1974, 35, 595-600.
[218] Kléman, M.; Williams, C. E. *J. Phys, Lett. (Paris)* 1974, 35, L-49 – l-51.
[219] Pershan, P. S. *J. Appl. Phys.* 1974, 45, 1590-1604.
[220] Williams, C. E.; Kléman, M. *J. Phys. Colloq. (Paris)* 1976, 36, C1-315 – C1-320.
[221] Williams, C. E.; Kléman, M. *Phil. Mag.* 1976, 33, 213-217.
[222] Bartolino, R.; Durand, G. *Mol. Cryst. Liq. Cryst.* 1977, 40, 117-132.
[223] Williams, C. E. *J. Phys. Colloq. (Paris)* 1978, 39, C2-46 – C2-57.
[224] Lejček, L. *Czech. J. Phys. B* 1982, 32, 767-771.
[225] Lejček, L. *Czech. J. Phys. B* 1990, 40, 1250-1259.
[226] Lejček, L.; Oswald, P. *J. Phys. II France* 1991, 1, 931-937.
[227] Benzekri, M.; Marcerou, J. P.; Nguyen, H. T.; Roullon, J. C. *Phys. Rev. B* 1990, 41, 9032- 9037.
[228] De Meyere, A. *Mol. Mat.* 1994, 3, 217-228.
[229] Holyst, R.; Oswald, P. *J. Phys. II France* 1995, 5, 1525-1532.
[230] Lelidis, I.; Kléman, M.; Martin, J. L. *Mol. Cryst. Liq. Cryst.* 1999, 330, 457-464.
[231] Brener, E. A.; Marchenko, V. I. *Phys. Rev. E* 1999, 59, R4752 – R4753.
[232] Slavinec, M.; Kralj, S.; Žumer, S. *Mol. Cryst. Liq. Cryst.* 2000, 351, 153-160.
[233] Braginsky, A. Yu. *Phys. Rev. B* 2003, 67, 174113 (6 pp).
[234] Rosenblatt, Ch. S.; Pindak, R.; Clark, N. A.; Meyer, R. B. *J. Phys. (Paris)* 1977, 38, 1105- 1115.
[235] Oswald, P.; Behar, J.; Kléman, M. *Phil. Mag.A* 1982, 46, 899-914.
[236] Clark. N. A.; Hurd, A. J. *J. Phys (Paris)* 1982, 43, 1159-1165.
[237] Takanishi, Y.; Ouchi, Y.; Takezoe, H.; Fukuda, A. *Jpn. J. Appl. Phys.* 1989, 28, L487-L489.
[238] Ouchi, Y.; Takanishi, Y.; Takezoe, H.; Fukuda, A. *Jpn. J. Appl. Phys.* 1989, 28, 2547-2551.
[239] Stewart, I. W. *Liq. Cryst. (UK)* 1993, 15, 859-869.
[240] Stewart, I. W.; Leslie, F. M.; Nakagawa, M. *Q. J. Mech. Appl. Math.* 1994, 47, 511-525.
[241] Kidney, K. J.; McKay, G.; Stewart, I. W. *Mol. Cryst. Liq. Cryst.* 2005, 438, 263-269.
[242] Williams, C. E. *Défauts de structure dans les sméctiques A* 1976, PhD Thesis, University of Paris, Orsay
[243] Urbach, W.; Boix, M.; Guyon, E. *Appl. Phys. Lett.* 1974, 25, 479-481.
[244] Hochbaum, A.; Labes, M. M. *J. Appl. Phys.* 1982, 53, 2998-3002.
[245] Rosenblatt, Ch, *Phys. Rev. Lett.* 1984, 53, 791-794.
[246] Sinha, Ch. P.; Rosenblatt, Ch.; Mirantsev, L.V. *Phys. Rev. E* 2002, 65, 041718 (4 pp).
[247] Meirovitch, E.; Freed, J. H. *J. Phys. Chem.* 1980, 84, 2459-2472.

[248] Känel, H. V.; Litster, J. D. Meligalis, J.; Smith, H. I. *Phys. Rev. A* 1981, 24, 2713-2719.
[249] Faetti, S.; Gatti, M; Palleshi, V.; Sluckin, T J. *Phys. Rev. Lett.* 1985, 55, 1681-1684.
[250] Mukai, S.; Nakagawa, M. *J. Phys. Soc. Jpn.* 1992, 61, 1560-1564.
[251] Poniewierski, A.; Holyst, R. *Phys. Rev. B* 1993, 47, 9840-9843.
[252] Li, Z.; Folks, W. R.; Lavrentovich, O. D. *Proc. SPIE* 1994, 2175, 21-32.
[253] Li, Z.; Lavrentovich, O. D. *Phys. Rev. Lett.* 1994, 73, 280-283.
[254] Hauck, G.; Koswig, H. D. *Mol. Cryst. Liq. Cryst.* 1991, 199, 159-165.
[255] Holyst, R. *Phys. Rev. Lett.* 1994, 72, 4097-4100.
[256] Fournier, J. B. *Phys. Rev. E* 1994, 50, 2868-2871.
[257] Evangelista, L. R.; Fontanini, S.; Malacarne, L. C.; Mendes, R.S. *Phys. Rev. E,* 1997, 55, R1279-R1281.
[258] Guo, W.; Bahr, Ch. *Phys. Rev. E* 2009, 79, 0117071 (5 pp).
[259] De Gennes, P. G. *C. R. Acad. Sc. Paris B* 1972, 275, 549-550.
[260] Kléman, M.; Parodi, O. *J. Phys.(Paris)* 1975, 36, 671-681.
[261] Kléman, M. *J. Phys. (Paris)* 1977, 38, 1511-1518.
[262] Holyst, R.; Poniewierski, A. *J. Phys. II France* 1993, 3, 177-182.
[263] Kléman, M.; Lavrentovich, O. D. *Phys. Rev. E* 2000, 61, 1574-1578.
[264] Khataevich, V. I.; Zeinally, A. Kh. *Pis'ma Zh. Tech. Fiz. (USSR)* 1979, 5, 651-654.
[265] Fournier, J. B.; Dozov, I.; Durand, G. *Phys. Rev. A* 1990, 41, 2252-2255.
[266] Rout, D. K.; Choudhary, R. N. P. *Mol. Cryst. Liq. Cryst.* 1989, 166, 75-90.
[267] Hareng, M.; Le Berre, S.; Thirant, L. *Appl. Phys. Lett.* 1974, 25, 683-685.
[268] Hareng, M.; Le Berre, S.; Metzger, J. J. *Appl. Phys. Lett.* 1975, 27, 575-576.
[269] Le Berre, S.; Hareng, M. *Ann. Phys (France)* 1978, 3, 327-330.
[270] Dunmur, D. A.; Walton, T. W. *Mol. Cryst. Liq. Cryst. Lett.* 1985, 2, 7-13.
[271] Findon, A.; Gleeson, H.; Lydon J. *Phys. Rev. E* 2000, 62, 5137-5142.
[272] Canter, S.; Crossland, W. A. *Electrical Commun. (G.B.)* 1986, 60, 87-93.
[273] De Gennes, P. G. *J. Phys. Colloq. (Paris)* 1969, 30, C4-65 – C4-71.
[274] Kahn, F. J. *Appl. Phys. Lett.* 1973, 22, 111-113.
[275] Sasaki, A.; Marioka, T.; Takagi, T.; Ishibashi, T. *IEEE Tr. El. Dev. 1975*, ED-22, 805-806.
[276] Sasaki, A.; Morioka, T.; Ishibashi, T.; Takagi, T. *Tech. Digest Int. El. Dev.* 1975, 401-403.
[277] Hareng, M.; Le Berre, S. *El. Lett.* 1975, 11, 73.
[278] Sasaki, A.; Morioka, T.; Ishibashi, T.; Takagi, T. *Proc. Sol. St. Devices* 1976, 15, 121-124.
[279] Hareng, M.; Le Berre, S., Hellen, R. *Rev. Techn. Thomson-CSF* 1977, 9, 373-394.
[280] Dewey, A. G.; Jacobs, J. T.; Huth, B. G. *Proc. S.I.D.* 1978, 19/1, 1-7.
[281] Hareng, M.; Le Berre, S. *J. Phys. Colloq. (Paris)* 1979, 40, C3-135 – C3-136.
[282] Sasaki, A.; Inoda, M., Ishibashi, T. *Proc. S.I.D.* 1980, 21/4, 341-344.
[283] Kahn, F. J.; Birecki, H. In *The Physics and Chemistry of Liquid crystal Devices*; Sprokel G. J.; Ed.; Plenum Publishing Corp.: New York, U.S., 1980, pp 79-93.
[284] Dewey, A, G. In *The Physics and Chemistry of Liquid crystal Devices* ; Sprokel G. J.; Ed.; Plenum Publishing Corp.: New York, U.S., 1980, pp 219-239.
[285] Crossland, W.; Ayliffe, P. J. *Proc. S.I.D.* 1982, 23, 9-13.
[286] Raynes, E. P. *Phil. Trans. R. Soc. London A* 1983, 309, 167-178.
[287] Birecki, H.; Naberhuis, S.; Kahn, F. J. *Proc. SPIE* 1983, 420, 194-199.

[288] Hochbaum, A. *Opi. Eng.* 1984, 23, 253-260.
[289] Dewey, A. G. *Opt. Eng.* 1984, 23, 230-240.
[290] Walker, C.J.; Crossland, W. A. *Displays* 1985, October, 207-211.
[291] Harrold, J.; Steele, C. *Proc. S.I.D.* 1985, 26/2, 141-146.
[292] Crossland, W. A.; Canter, S. *Electronic Eng.* 1985, 57, 35-41.
[293] Sasaki, A. *Mol. Cryst. Liq. Cryst.* 1986, 139,103-121.
[294] Steele, C. A.; Harrold, J.; Stanton, J. P.; Daley, R. *Proc. SPIE* 1987, 760, 70-73.
[295] Nakamuram K.; Oda, Y.; Sigimoto, N. *J. Appl. Phys.* 1986, 59, 593-595.
[296] Nagae, Y.; Kaneko, E.; Mori, Y.; Kawakami, H.; Hashimoto, T. *Proc. S.I.D.* 1987, 28/1, 55-58.
[297] Nakamura, K.; Nanataki, H.; Arimoto, M. *J. Appl. Phys.* 1990, 67, 996-1000.
[298] Kreuzer, N.; Tschudi, T.; De Jeu, W. H.; Eidenschink, R. *Appl. Phys. Lett.* 1993, 62, 1712-1714.
[299] Dau, N.; Qimude, Sh.; Nawa, N. *Chinese Sci. Bull.* 1991, 36, 1521-1534.
[300] Nawa, N. *Mol. Cryst. Liq. Cryst.* 1991, 196, 39-46.
[301] Nawa, N *Mol. Cryst. Liq. Cryst.* 1991, 208, 77-83.
[302] Armitage, D, *J. Appl. Phys.* 1981, 52, 1294-1300.
[303] Armitage, D, *J. Appl. Phys.* 1981, 52, 4843-4851.
[304] Wilson, T.; Boyd, G. D.; Westerwick, E, H. *IEEE Tr. El. Dev.* 1983, ED-30, 245-247.
[305] Hareng, M.; Le Berre, S.; Mourey, B.; Mouton, P.; Perbet, J.-N.; Thirant, L. *Proc, S.I.D.*1983, 24/2, 167-173.
[306] Lee, M. H.; Depp, S. W.; Juliana, A.; Moore, J. O.; Koepcke, R. W. *J. Appl. Phys.* 1985, 57,1055-1062.
[307] Chu, W. H.; Jacobs, J. T. *IBM J. Res. Dev.* 1978,22, 40-50.
[308] Chu, W. H. ; Yoon, D. Y. *Mol. Cryst. Liq. Cryst.* 1979, 54, 245-260.
[309] Nawa, N.; Nuo, D. *Jpn. J. Appl. Phys.* 1991, 30, 1038-1042.
[310] Tani, C. *Appl. Phys. Lett.* 1971, 19, 241-242.
[311] Smit, K. *Electronika (USSR)* 1985, N 15, 13-14.
[312] Jákli, J.; Saupe, A. Mol. Cryst. Liq. Cryst. 1992, 222, 101-109.
[313] Prost, J.; Toner, J. *Phys. Rev. A* 1987, 36, 5008-5014.
[314] Durand, G. *C. R. Acad. Sc. Paris B* 1972, 275, 629-631.
[315] Clark, N. A.; Meyer, R. B. *Appl. Phys. Lett.* 1973, 22, 493-494.
[316] Ribotta, R. *C. R. Acad. Sc. Paris B* 1974, 279, 295-296.
[317] Clark, N. A. *Phys. Rev. A* 1976, 14, 1551-1554.
[318] Oswald, P.; Ben-Abraham, S. I. *J. Phys.(Paris)* 1982, 43, 1193-1197.
[319] Bevilacqua, G.; Napoli, G. *Phys. Rev. E* 2005, 72, 041708 (5 pp).
[320] Stewart, F.: Stewart, I. W. *Mol. Cryst. Liq. Cryst.* 2007, 478, 23-32.
[321] Hartmann, W. *The Surface Stabilized Ferroelectric LC Structures with its Bistability* Ph. D. Thesis, Philips Res. Laboratories, Eindhofen, The Netherlans 1990, pp 105-113.
[322] Vertogen G.; De Jeu, W. H. *Thermotropic Liquid crystals. Fundamentals*, Springer Verlag: Berlin, 1988, chap. 5.
[323] Durand G. In *Cristalli Liquidi 1^AScuola Nationale UNICAL*, Italy 1981, pp 525-542.
[324] Orsay Group on Liquid crystals *J. Phys. Colloq. (Paris)* 1975, 36. C1-305 – C1-313.
[325] Hinov, H. P. *A method for preparation of liquid crystal cell with storage*, Certificat for Authorship, N 33396, Reg. N 48390 1980, 1-8.

[326] Hinov, H. P. *A method for preparation of liquid crystal cell,* Certificat for Authorship, N 42494, Reg. N 76021 1986, 1-9.
[327] Osipov, M. A.; Sambles, J. R.; Ruan, L. *Liq. Cryst. (UK)* 2003, 30, 823-830.
[328] Ruan, L. Z.; Osipov, M. A.; Sambles, J. R. *Phys. Rev. Lett.* 2001, 86, 4548-4551.
[329] Ruan, L. Z.; Sambles, J. R.; Stewart, I. W. *Phys. Rev. Lett.* 2003, 91, 033901 (4 pp).
[330] Pishnyak, O. P.; Nastishin, Yu. A.; Lavrentovich, O. D. *Phys. Rev. Lett.* 2004, 93, 109408.

Chapter 2

FORMATION OF SMECTIC-C LIQUID CRYSTAL FROM A NON-DEFORMED, WEAKLY-DEFORMED AND STRONGLY-DEFORMED NEMATIC LAYER. POSSIBILITY FOR THE DETERMINATION OF CERTAIN SMECTIC-C ELASTIC CONSTANTS

2.1. GENERAL INFORMATION FOR THE STRUCTURE AND THE ELASTIC ENERGY OF THE SMECTIC-C

The molecules of the smectic-C, in contrast to the smectic-A, are inclined with respect to the smectic-C layers' normal. The mean orientation of an ensemble of smectic-C molecules is represented by a smectic-C director which is designated as in the case of nematics and cholesterics, by the director **n**. For this reason, the structure of the smectic-C as a whole has a two-fold (biaxial) symmetry of a monoclinic class with a plane of symmetry as is shown in Figure 2.1.

An unit vector **c** is introduced which shows the preferable mean orientation of the projection of one ensemble of smectic-C molecules on the basic plane (x, y). This is the well-known C2 director. The angle ω which is described usually by θ is temperature dependent [1]. Some smectic-C liquid crystals, however, have a temperature independent tilt.

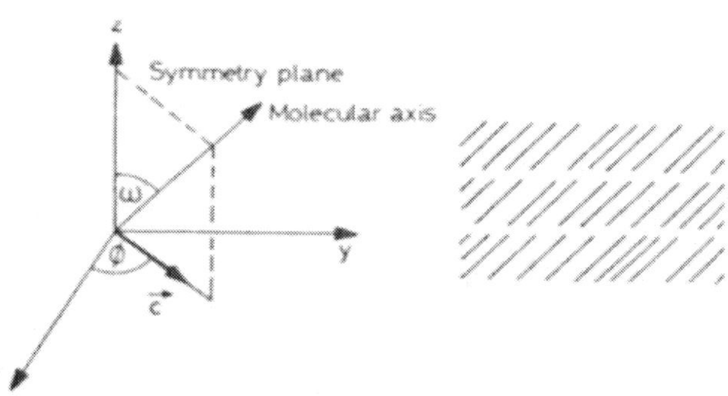

Figure 2.1. The structure of smectic-C.

The original experimental results obtained for the smectic-C which will be exposed here, are connected with undulations and dislocations which show the minimization of the elastic energy of the smectic-C under external influences. The elastic energy of the smectic-C consists of three parts. The first one had been obtained for the first time by Saupe [2] who included all the possible reorientations of the smectic-C director around the normal to the smectic-C plane:

$$F_{c_1} = (1/2)B_1\left(\frac{\partial\varphi}{\partial x}\right)^2 + (1/2)B_2\left(\frac{\partial\varphi}{\partial y}\right)^2 + (1/2)B_3\left(\frac{\partial\varphi}{\partial z}\right)^2 + B_{13}\frac{\partial\varphi}{\partial x}\cdot\frac{\partial\varphi}{\partial z} \tag{2.1}$$

where φ is the angle between the projection of the director on the smectic-C plane, B_1, B_2 and B_{13} are the elastic constants describing the elastic resistance of the smectic-C to all possible reorientations of the smectic-C director. The second part of the smectic-C elastic energy, obtained for the first time by the Orsay Liquid crystal Group [3] includes the elasticity of the smectic-C connected with the bending of the smectic-C layers as well as with the dilation or contraction of the smectic-C layers leading a change of the distance between every two smectic-C layers:

$$\begin{gathered} F_{c_2} = (1/2)A_{11}\left(\frac{\partial^2 u}{\partial x\partial y}\right)^2 + (1/2)A_{12}\left(\frac{\partial^2 u}{\partial x^2}\right)^2 + (1/2)A_{21}\left(\frac{\partial^2 u}{\partial y^2}\right)^2 \\ + (1/2)\overline{B}\left\{\frac{\partial u}{\partial z} - (1/2)\left[\left(\frac{\partial u}{\partial x}\right)^2 + \left(\frac{\partial u}{\partial y}\right)^2\right]\right\} \end{gathered} \tag{2.2}$$

where u is the displacement of the smectic-C layers along the axis Z, i.e., perpendicular to the smectic-C planes, A_{11}, A_{12} and A_{21} are the elastic constants of the smectic-C which resists to the bending of the smectic-C layers and $\overline{B}$ is the contraction (dilation) constant of the smectic-C layers. The third of the smectic-C elastic energy has been also obtained by the Orsay Liquid crystal Group [3]. It describes the connection between the bending of the smectic-C layers and the change in the orientation of the directors **n** and **c:**

$$F_3 = C_1\frac{\partial^2 u}{\partial x\partial y}\frac{\partial\varphi}{\partial x} + C_2\frac{\partial^2 u}{\partial y^2}\frac{\partial\varphi}{\partial y} \tag{2.3}$$

Furthermore, relations and inequalities have been obtained between the elastic constants, based on the stability conditions:

$$\begin{array}{ll} A_{11}^2 - 4A_{21}A_{12} < 0 & A_{12}, A_{21} > 0 \\ B_1B_3 - B_{13}^2 > 0 & B_1, B_2, B_3 > 0,\ \ \overline{B} > 0 \\ A_{12}B_1 - C_1^2 > 0 & A_{21}B_2 - C_2^2 > 0 \end{array} \tag{2.4}$$

Rapini [4] was the first author who presented the elastic energy of the smectic-C in terms of mathematical operations on three independent unit vectors. Much later, Leslie et al., rerwote the smectic-C elastic energy obtained by Rapini with the aid of new symbols [5, 6]. All the smectic-C elastic constants, with exception of the elastic constant $\overline{B}$, are similar to those of the nematics and cholesterics They have obtained the following form of the elastic energy of the smectic-C:

$$\begin{aligned} W &= \frac{1}{2}A_{12}\left(\vec{b}.\nabla x\vec{c}\right)^2 + \frac{1}{2}A_{21}\left(\vec{c}.\nabla x\vec{b}\right)^2 + A_{11}\left(\vec{b}.\nabla x\vec{c}\right)\left(\vec{c}.\nabla x\vec{b}\right) + \frac{1}{2}B_1\left(\nabla.\vec{b}\right)^2 \\ &+ \frac{1}{2}B_2\left(\nabla.\vec{c}\right)^2 + \frac{1}{2}B_3\left[1/2\left(\vec{b}.\nabla x\vec{b} + \vec{c}.\nabla x\vec{c}\right)\right]^2 + B_{13}\left(\nabla.\vec{b}\right)\left(1/2\,\vec{b}.\nabla x\vec{b} + \vec{c}.\nabla x\vec{c}\right) \\ &+ C_1\left(\nabla.\vec{c}\right)\left(\vec{b}.\nabla x\vec{c}\right) + C_2\left(\nabla.\vec{c}\right)\left(\vec{c}.\nabla x\vec{b}\right) \end{aligned} \tag{2.5}$$

The comparison of (2.4) and (2.5) clearly shows that one of the inequalities from the relations (2.4) is replaced by:

$$A_{12}B_2 - C_1^2 > 0 \tag{2.6}$$

New inequalities have also been obtained which must be added to the inequalities expressed by relations (2.4) and (2.6) [7]:

$$\begin{aligned} &A_{12} + A_{21} + 2A_{11} > 0,\ A_{12} + A_{21} - 2A_{11} > 0 \ \text{или}\ \left|A_{11}\right| < (1/2)(A_{12} + A_{21}) \\ &B_1 + B_3 \pm 2B_{13} > 0 \ \text{или}\ \left|B_{13}\right| < \frac{1}{2}(B_1 + B_3) \\ &B_2 + A_{12} \pm 2C_1 > 0 \ \text{или}\ \left|C_1\right| < \frac{1}{2}(B_2 + A_{12}) \\ &B_2 + A_{21} \pm 2C_2 > 0 \ \text{или}\ \left|C_2\right| < \frac{1}{2}(B_2 + A_{21}) \end{aligned} \tag{2.7}$$

All the smectic-C elastic constants, with exception of the elastic constant $\overline{B}$, are of one and the same order of 10^{-6} dyn. The value of the smectic-C elastic constants can be obtained, for example, by the Fréedericksz transition in planar or non-planar: spherical-, wedge-shaped or cylindrical geometries [4, 7-13]. The theoretical and experimental results permitted to obtain some of the elastic constants of the smectic-C, designated by A in References [9] and [10] as well as in References [14] and [15]. For instance, Meirovitch et al., suggested that the smectic-C director moves in a step-like manner along the cone surface. Previously, Pelzl et al., have observed such a Fréedericksz transition in their first experimental studies on the smectic-C [16, 17]. Schiller and Pelzl have estimated the mean value of the elastic constants of the smectic-C $B_1 \sim B_2$ to be in the range of 10^{-6} dyn [18]. Various smectic-C elastic constants have been measured by Galerne et al., [19] and by Pelzl et al., [20]. It is interesting that Pelzl et al., have determined the rotational viscosity of the smectic-C phase, designated by C_6, and proved that it is from 20 to 25 times smaller relative to the rotational viscosity of the nematic phase of the same liquid crystal. Dierking et al., [21] have measured the value of the constant B, the reduced elastic constant B_0 as well as the rotational viscosity of the liquid crystals from C1 to C5 with a nematic-smectic-C phase transition (C3, C4) or with a nemati-smecticA-smectic-C phase transitions (C1, C3 and C5) (see also the results, presented in Reference [20]). Depending on the temperature and the kind of the liquid crystal, they have obtained for the elastic constant B the value between $38x10^{-7}$ and $3x10^{-7}$ dyn. Brown et al., [22] re-measured the value of $B_1$and B_2 for the liquid crystal SCE8R and obtained more accurately, in their opinion values. On the other hand, in a previous paper had been estimated and the value of $B_{13}$of some liquid crystal to be between, 10^{-8} and 10^{-7} dyn [23]. Shortly, the experimental measurements of various authors had shown that B_1 and B_2 are of the same order of magnitude and B_{13} is of one order of magnitude smaller in comparison with the case of nematics [24]

2.2. A Basic Information About the Nematic-Smectic-C Phase Transition

This phase transition has been theoretically considered by de Gennes [25], Chen and Lubensky [26] and Chu and McMillan [27], and also by other authors [28-30]. For instance, Chen and Lubensky have obtained the following values for the increase of the elastic constant of splay, twist and bend:

$$\delta K_1 = \frac{9kT}{64\pi} q_\perp \left(q_{II}^2 \frac{\xi_\perp^3}{\xi_{II}} + \frac{2}{9} q_\perp^2 \xi_{II} \xi_\perp \right)$$

$$\delta K_2 = \frac{3kT}{64\pi} q_\perp \left(q_{II}^2 \frac{\xi_\perp^3}{\xi_{II}} + \frac{2}{9} q_\perp^2 \xi_{II} \xi_\perp \right) \quad \delta K_3 = \frac{kT}{24\pi} q_{II}^2 q_\perp \xi_{II} \xi_\perp \tag{2.8}$$

where $q_\parallel$ and $q_\perp$ are the components of the vector of the wave density of the smectic- phase, $\xi_\parallel$ and $\xi_\perp$ are the longitudinal and transversal correlation lengths of the smectic-C order, i.e., they are connected with the size of the cybotactic groups [31-33]. The mean-field approximation of the theory based on the theory of Landau-Ginzburg, points out towards a second order nematic-smectic-C phase transition. The inclusion of the fluctuations of the director near the temperature of the transition however, leads to a change in the phase transition from a second order to a first order, as in the case of the nematic-smectic-A phase transition (the well- known effect of Halperin-Lubensky-Ma). Let us recall that the theory of de Gennes predicted that the three elastic constants of Frank, diverge as $\xi^{3/2}$, where ξ is the smectic-C correlation lengt, the theory of Chen and Lubensky predicted a divergence of the same elastic constants as ξ^2, and the theory of Chu and McMillan predicted a divergence as ξ.

2.3. Formation of the Smectic-C Phase from Strongly-Deformed Nematic Surface Layer in the Presence of Electric Field: General Ideas and Formulation of the Problem

The A-, B-, C-, etc. smectics with a preceding nematic phase can be formed from a nematic which is deformed in the boundary regions and is oriented in the middle part of the liquid crystal cell smectic-C with a nematic-smectic-C phase transition of second or of a slightly-first order. It can be formed first in the middle parts of the cells where the nematic is oriented and later in the boundary regions since in these regions the nematic-smectic-C phase transitional temperature decreases as in the smectic-A of a first kind. This way of the smectic-C growth permits the formation of new smectic-C textures, their observation and the better study of the properties of the smectic-C. Furthermore, the formation of the smectic-C phase in this way can reveal also important information about the strength of the surface interactions either quantitatively or qualitatively, in both the smectic-C phase and the nematic phase. Up to now these surface interactions have been insufficiently studied for the smectic-C phase.

We will initially consider the simpler case of the smectic-C formation of the liquid crystal NOBA (p-n-nonyloxybenzoic acid) from variously oriented and non-deformed nematic films. We will present the formation of the smectic-C phase after rapid cooling of electrically-deformed planar, hybrid (homeo-planar) and twisted NOBA films. During the experimental investigations a number of either well known or novel smectic-C textures are obtained giving important information about the possible matching of both the non-deformed and strongly-deformed smectic-C, one of which is nucleated in the middle part of the samples, and the other one is formed in the boundary regions. The possible ways of the smectic-C growing, including the orientation of the smectic-C directors and the position of the smectic-C planes, were clarified as well as the crucial role of the initial twist in the nematic phase and the strength of the voltage, applied across both the nematic and smectic-C phase for the creation of various smectic-C textures were demonstrated. Moreover, the possible role of the surface forces and the possibility for estimation of some elastic constants of the smectic-C have been discussed.

2.4. Formation of the Smectic-C Phase from Non-Deformed Planar, Homeotropic and Twisted Nematic NOBA Films as well as from Electrically-Deformed Planar and Twisted NOBA Films: Preliminary Remarks

The nematic-smectic-C phase transition is accompanied by a relatively small enthalpy measured to be in the range of 0.4 kcal/mol [34]. Rao and Pisipati [35] have considered this phase transition for NOBA and have considered that it is of a first order. The characteristic tilt of the director with respect to the smectic-C planes is of the order of 45^0 and is temperature independent [36]. The electrical properties of NOBA have been studied by Chou and Carr [37-39] and by Kuvatov et al., [40-42]. Their experimental results showed that the dielectric anisotropy is positive in the nematic phase and negative in the smectic-C phase. They also showed that the temperature change in both phases is weak, except around the nematic-smectic-C phase transition. The value of the dielectric anisotropy in the nematic and the smectic-C phases has been measured to be + 0.028 and – 0.025, respectively. Furthermore, these authors proved also that the biaxiality of the dielectric tensor of NOBA in the smectic-C phase is weak, i.e., that NOBA in the the smectic-C is practically uniaxial.

The liquid crystal NOBA can be represented by the following chemical formula and the following phases and phase transitional temperatures:

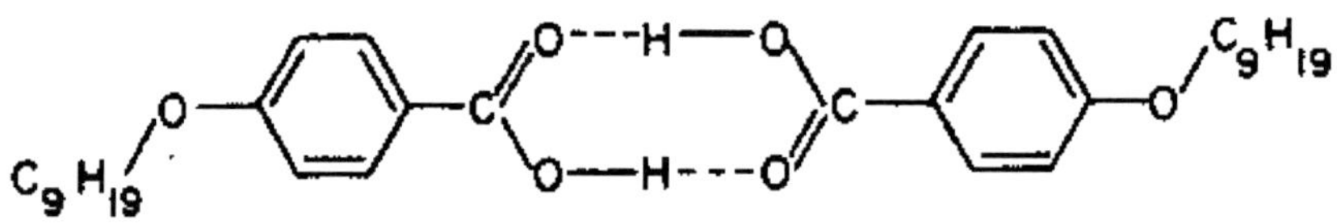

$$NOBA: I_{143°C} - N_{117°C} - S_m C_{94°C} - C$$

It was introduced between two NESA coated semitransparent glass plates separated by polymer spacers 10 to 50 μm thick. We did not use in our experiment a thermostatic plate and a thermo-regulator. Since the construction of the cell was held by two massive metal holders (see Figure 1.1in chapter 1), they themselves serve as warming elements after heating of the cell up to the highest temperature of 150^0 C. In this way, one notes a strong temperature gradient in the electrode planes, directed from the lateral metal holders to the middle part of the cell. This permits easy observation of all the pre-transitional effects around the nematic-smectic-C phase transition, the study of a large area (several cm^2) of the liquid crystal and easy photography of the observed liquid crystal structures with objectives having small focal lengths, with magnification of x400. The temperature distribution in such cells is discussed earlier for the case of the nematic-smectic-A phase transition. The prepared liquid crystal cells were studied in a free cooling regime. The temperature gradient was from the lateral metal holders to the middle part of the cell. On the other hand, the temperature was homogeneous in the perpendicular direction, which coincides with the direction of the applied electric field and with the direction of observation of the cells under the polarizing microscope. The treatment of the glass plates by SiO under vacuum evaporation determines planar or twist nematic layer with a big interfacial surface energy of interaction between the substrate and the liquid crystal, which is of the order of erg/cm^2. The homeotropic orientation of the nematic was determined by preliminary treatment of the glass plates with Lecithin ensuring in this way a surface energy of two orders of magnitude smaller, which is in the range of 10^{-2} erg/cm^2. The voltage has a frequency of 10 kHz and a value between U_c and $15U_c$, where U_c is the threshold voltage between 26-30 V. The voltage was applied on the nematic and the smectic-C phases. It was removed after the formation of the smectic-C textures.

2.5. Microscopic Observations: Formation of Smectic-C Phase from Oriented and Electrically Deformed Planar and Twisted Nematic Films

a. Fast Cooling of Thin Planar Films of NOBA with a Thickness of 10-26 μm to the Smectic-C Phase

This manner of cooling often led to the formation of many local smectic-C monodomains divided by π-inverse walls [43]. Pre-transitional stripes, similar to those investigated by Cladis and Torza in the smectics-A, observed for the first time by Arora et al., [44] and noted in the book by Demus and Richter [43]. The pre-transitional stripes were observed in cells with a thickness above 10-15 μm which shows that their period was determined from the given temperature near to the nematic-smectic-C phase transition. Let us mention that their shape follows the large temperature gradient, being formed in the electrode plane. Smectic-C blurred Schlieren textures were formed in cells with a thickness of 50 μm, which were not controlled by the surface interactions of the liquid crystal [43].

b. Fast Cooling of Thin Twisted Nematic NOBA Films with a Thickness of 10-26 μm to the Smectic-C Phase

This manner of cooling of electrically-realigned thin planar or twisted nematic NOBA films, in the presence of the applied voltage with an above-threshold amplitude led to observation of oriented liquid crystal in the middle part of the liquid crystal layer and strongly-deformed boundary regions and finally, to the formation of smectic-C Schlieren texture with one-dimensional wave-like instability of Fergason-Johnson-Saupe [46] at a lower voltage between 20 and 60 V and a frequency of 20 kHz. Here we must note that undulation instability in the smectic-C was also observed by Bartolino and Durand [47], Kumar [48], Furnier and Galatola [49], and Torgova et al., [50]. These domains are shown in Figure 2.2.

At a further decrease of the temperature in the smectic-C phase, another undulation of the smectic-C layer appears, superimposed on the first undulation, in a perpendicular direction, shown in Figure 2.3.

Let us mention that the one-dimensional undulation instability in the smectic-C phase appeared at a low voltage between 20 and 60 $V_{r.m.s.}$ and a frequency of 20 kHz (the threshold voltage for untwisting the layers in the nematic phase was between 16 and 20 $V_{r.m.s.}$, depending on the impurities in the liquid crystal). Second, the voltage in the smectic-C phase was not capable to change significantly the smectic-C layer undulations. One sees also smectic-C local monocrystals, in part decorated with edge dislocations, which appear at a higher voltage of 100 $V_{r.m.s.}$ (Figure 2.4).

In addition, at an increase of the voltage up to 280 $V_{r.m.s.}$, we saw smectic-C local monocrystals and smectic-C local monocrystals decorated with considerably depressed in their amplitude edge dislocations (Figure 2.5).

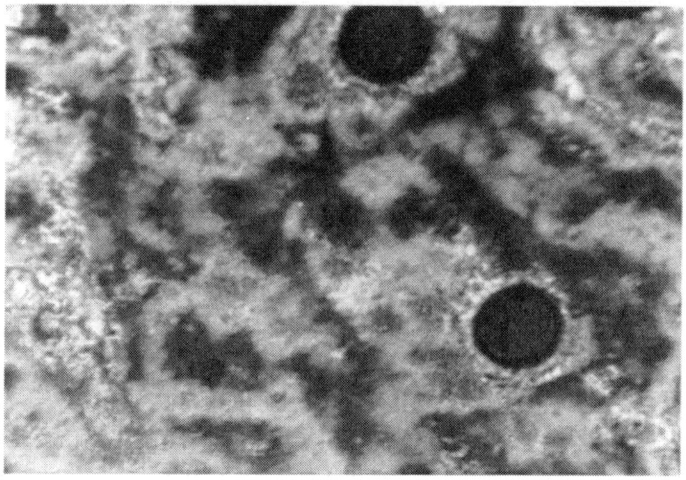

Figure 2.2. Smectic-C Schlieren texture with smectic-C layer undulation instabilities in a thin (26 μm) NOBA films produced by second-order effects after rapid cooling of the twisted nematic phase. Crossed nicols, the short side of the photo corresponds to 480 μm.

Figure 2.3. Smectic-C Schlieren texture with two superimposed smectic-C layer undulation instabilities (cross-hatched patterns) in a thin (26 μm) NOBA film produced by first and second-order effects after rapid cooling of the twisted nematic phase. Crossed nicols, each side of the photo corresponds to 510 μm.

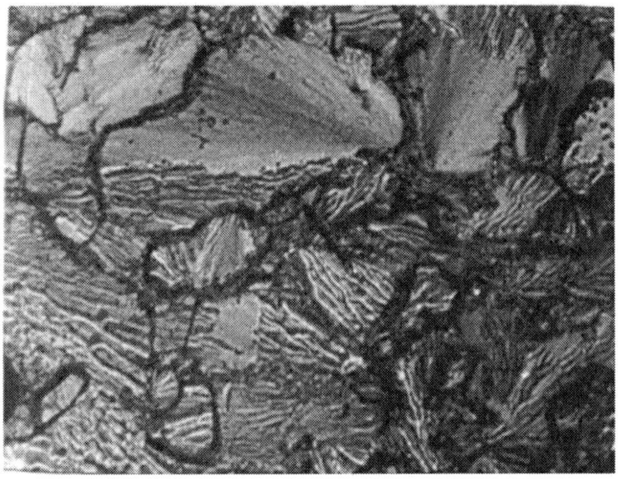

Figure 2.4. Smectic-C local monocristals with tilted inversion walls and edge dislocation lines after rapid cooling of electrically deformed twisted NOBA film with a thickness of 26 μm. Parallel nicols, U_{eff}=220$V_{r.m.s.}$, f=10 kHz, the short side of the photo corresponds to 330 μm.

Figure 2.5. Smectic-C local monocristals with edge dislocation lines obtained after rapid cooling of electrically deformed twisted NOBA film with a thickness of 26 μm. parallel nicols, U_{eff}=280$V_{r.m.s.}$, f=10 kHz, the short side of the photo corresponds to 330 μm.

The further increase of the voltage to the limit value (in our experiment) of 300 $V_{r.m.s.}$ led to significant decrease of the period and amplitude of the edge-dislocations in the smectic-C phase. The crude measurements of their period clearly show the 1/E law, where E is the strength of the applied electric field across both the nematic and the smectic-C phases (Figure 2.6).

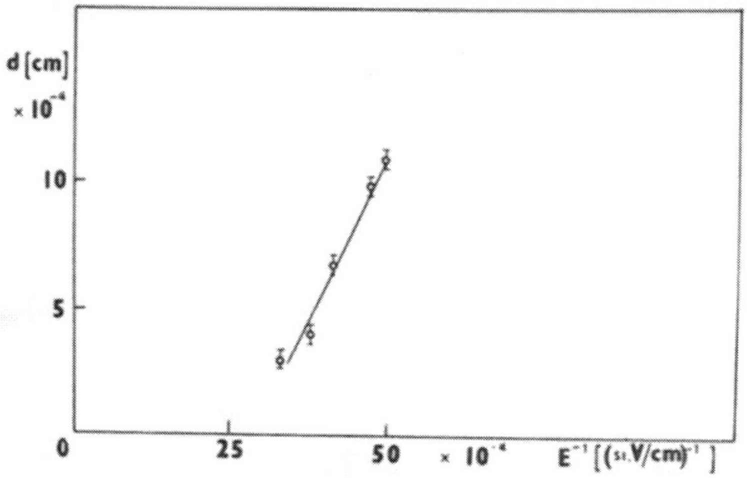

Figure 2.6. The period of the edge dislocations (in cm) vs the inverse strength of the electric field (in $(stV/cm)^{-1}$) of the applied voltage.

Furthermore, we observed the decrease in the period of the dislocations by jumps when the voltage was increased. Finally, the edge dislocations were preserved embedded into the local smectic-C regions at removal of the applied voltage.

2.6. Stripes of Fergason-Johnson-Saupe: Influence of the Thickness of the Liquid Crystal Films under Study, the Initial Orientation of the Nematic Phase and the Surface Anchoring of the Smectic-C Phase

All the smectic-C layer undulations observed up to now have been obtained after a rather rapid cooling of the preceding nematic or smectic-A films with various thicknesses, determined by Teflon or other hard-compressible spacers down to the smectic-C phase. Such stripes have not been observed in a smectic-C with one oriented phase since the characteristic smectic-C tilt in this case is temperature-independent. The stripes of Fergason-Johnson-Saupe can be easily observed in the smectic-C phase with preceding smectic-A phase since the tilt of the director in the smectic-C phase is changed with decrease of the temperature which in turn leads to a change in the thickness of the smectic-C layer. All the stripes of Fergason-Johnson-Saupe have been observed in Schlieren textures, characteristic both of the nematic or smectic-C phases. We have experimentally obtained the stripes of Fergason-Johnson-Saupe for the first time for liquid crystals with nematic and smectic-C phases and temperature independent tilt of the smectic-C director. They were observed in Schlieren textures with disclinations of ±1, ±2, etc. which are perpendicular to the glass plates and π-inversion walls dividing the adjacent local smectic-C domains. Our experimental observations clearly show that waving of the smectic-C layers cannot be obtained for the case of vertical and tilted smectic-C layers.

It is well known from the literature that the strong θ-polar and φ-azimuthal anchoring of the smectic-C directors at the boundaries achieved, for instance, by means of thin SiO films, deposited under vacuum evaporation, can govern the formation of local smectic-C monodomains with vertical or oblique orientation of the smectic-C relative to the glass plates (planar position of the smectic-C molecules). (Figure 2.7).

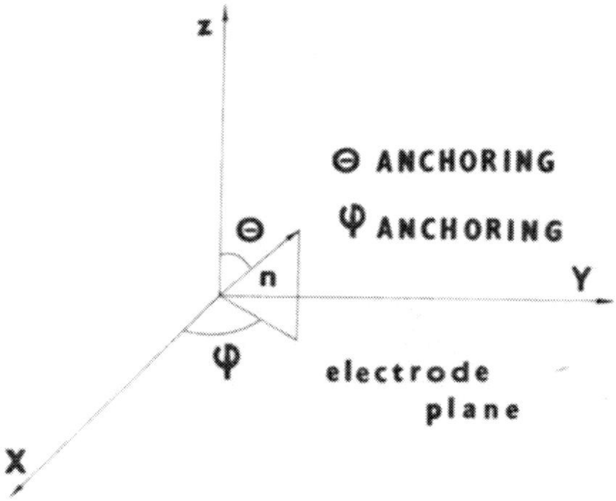

Figure 2.7. Schematic representation of the anchoring of the smectic-C.

Our experimental observations clearly show that smectic-C layer undulations are not possible for the case of either vertical or oblique position of the smectic-C planes relative to the glass plates (Figure 2.8).

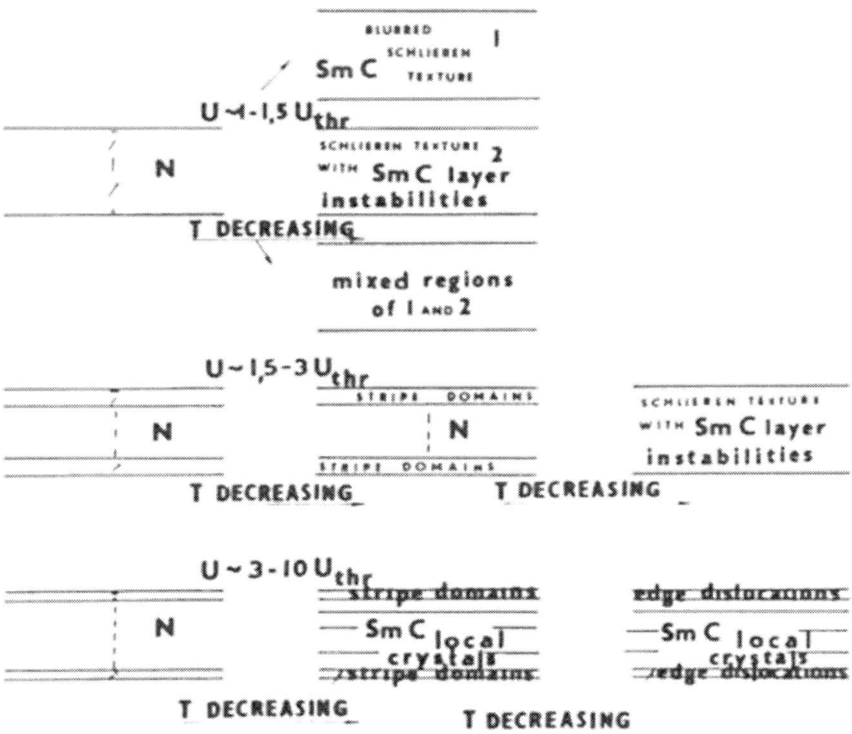

Figure 2.8. Schematic representation of various nematic and smectic-C textures obtained before and after rapid cooling of twisted nematic NOBA films, resp.

The explanation is very simple: when the smectic-C layers are parallel to the glass plates, they are very long and can be easily deformed. Inversely, when they are perpendicular to the glass plates, they are very short and could not be deformed. The experimental investigation of Price and Bak [51] unambiguously pointed out 20-50 μm penetration depth of the short-pitch cholesteric or smectic-A surface forces into the films under study which is much less than the penetration of surface forces into large-pitch cholesterics and nematics. It seems that this small penetration depth of the surface forces into smectic phases is due not to their weakness [52, 53], but rather to the very bulk forces of the smectics which are able to overcome the surface forces even for relatively thin smectic films (see, for instance, the remarkable paper of Durand for the penetration of the smectic-A layer undulations [54]). We verified these ideas with a nearly homeotropic nematic layer with a small tilt of the director. Such orientation can be obtained after rubbing of the glass plates, for instance with abrasive paste, and covering them with a surfactant – Lecithin to ensure the nearly homeotropic alignment. The nematic after this procedure was cooled down to the smectic-C phase. In this way, very well oriented stripe domains were obtained in the liquid crystal cells. These domains are parallel to the C2 axis of the smectic revealing effects of a second order, i.e., they are formed by strong connection between the bent layers and the tilt of the smectic-C molecules in ensemble. It is known that these stripes appear first in the vicinity of to the smectic-C – nematic phase transition. The perpendicular stripes are of first order and do not show such a coupling and can rarely appear at lower temperature in the smectic-C phase as well [46]. (Figure 2.9 and Figure 2.10).

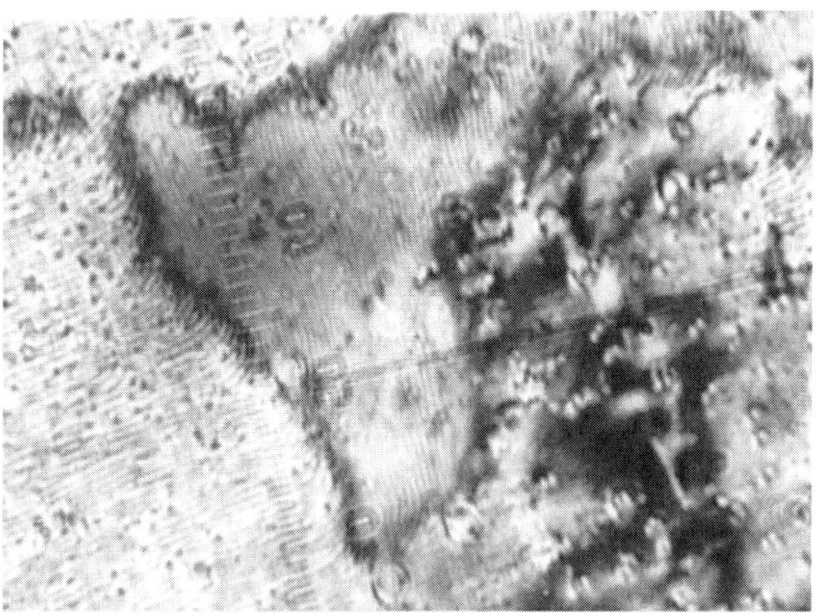

Figure 2.9. Formation of Schlieren smectic-C texture consisting of $_{\Pi}$-inversion bulk walls with pronounced smectic-C layer undulations after rapid cooling of a thin (26 μm) homeotropic nematic NOBA film, P45°A. The short side of the photo corresponds to 330 μm.

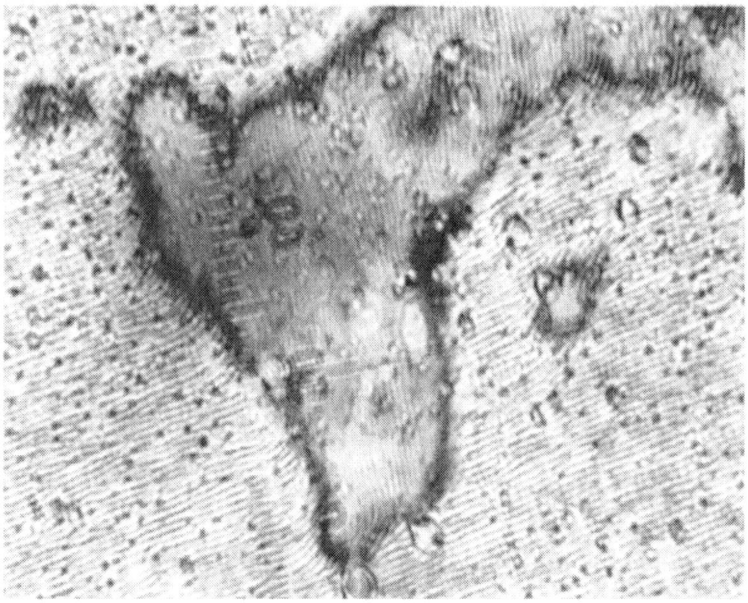

Figure 2.10. The same image from Figure 2.9 after 15 s.

2.7. The Stripes of Fergason-Johnson-Saupe: Possibilities for the Estimation of Certain Material Constants of the Smectic-C

The basic formulae in the theory of Johnson and Saupe [46] are:

$$\Lambda_1 = \left(\frac{A_{12}}{\overline{B}} \right)^{1/2} \tag{2.9}$$

$$\Lambda_2 = \left(A_{21} - \frac{C_2^2}{B_2} \right)^{1/2} \left(\overline{B}\right)^{-1/2} \tag{2.10}$$

$$A_{12} = \frac{\lambda_{c_{1,2}}^2}{4\pi d} \tag{2.11}$$

The comparison of these formulae determining the penetration depth of the domains of first order (2.9), domains of second order (2.10) and their connection with the period of the domains and the thickness of the liquid crystal cells (2.11) to our experimental results permitted, first, to estimate the modulus $\overline{B}$ of the smectic-C layers for the case of the NOBA liquid crystal:

$$\overline{B} \sim 3x10^{+4}\ erg/cm^3,\ if\ A_{12} = 10^{-5}\ dyne$$

(see also the estimations of Hinov and Petrov [55]). Second, these formulae permitted to estimate the correlation between some of the elastic constants of the smectic-C phase of NOBA, which can be obtained from the comparison of the widths of the two kinds of domains:

$$A_{12} = 16\left(A_{21} - \frac{C_2^{\ 2}}{B_2}\right) \tag{2.12}$$

It is clear also that increase in the value of A_{12} with one order of magnitude would rise also the value of $\overline{B}$. On the other hand, the observation of "broken" focal conics in the smectic-C phase of NOBA at lower temperatures [56] shows the following inequality:

$$A_{12} >> A_{21} \tag{2.13}$$

It is valid for the case of the liquid crystal NOBA at those temperatures (see the calculations of Kléman in [57, t II, pp 41-42]). On the other hand, from (2.13) one can obtain for this concrete case the following inequality:

$$A_{21} >> C_2^2 / B_2 \tag{2.14}$$

When the constant C2 is vanishing (C2 = 0) which is fulfilled in our experiment [55], $A_{12} = 16A_{21}$ and $|A_{11}| < (1/8)A_{12}$. We will finish this part of the work with two comments about the kind of the stripes and about the value of the modulus $\overline{B}$. In some cases, the optical images of the stripes of Fergason-Johnson-Saupe and the stripes of Findon-Gleeson-Lydon [58] are very alike. However, there is a great difference between them: undulations of the smectic-C layers versus arrays of chevrons with zig-zag orientation of the smectic-C layers, The bright and dark spots in the donains of Findon-Gleeson-Lydon can be explained by lines of interruptions while the bright and dark spots in the domains of Fergason-Johnson-Saupe can be explained in two ways: first, the undulations can be replaced by a system of confocal domains as observed by Fournier and Galatola [49] (see also the References from [59] to [61]). The second explanation is the existence of two superimposed undulations. The second comment concerns the value of the modulus $\overline{B}$ of the smectic-C with a preceding nematic phase. This elastic modulus has been rarely measured and the difference in the measured values is sometime of several orders of magnitude. [62-64] There is no doubt that this modulus for smectic-C with a preceding nematic phase is smaller several orders of magnitude [46, 55] in comparison with smectic-C with a preceding smectic-A phase. This fact has been theoretically pointed out by Johnson and Saupe [46] and by Wulf [65].

2.8. Edge Dislocations in Smectic-C: Conjectures and a Growth Model

The orientation of the smectic-C monocrystals, with or without edge dislocations, is in all possible directions, which shows that they are formed in the middle part of the liquid crystal cells where the influence of the boundaries is not felt. This way of the local smectic-C monocrystal growth confirms the finding for the case of the smectic-A that the nematic – smectic-C phase transitional temperature decreases when a smectic-C with a second or a slightly first nematic – smectic-C phase transitional order is formed from a strongly-deformed preceding nematic phase (let us recall that the enthalpy of the nematic-smectic-C phase transition is smaller relative to kT).

Lefévre et al., [66] have observed for the first time in 1971 edge dislocations in the smectic-C phase of the liquid crystal DOBCP with a nematic-smectic-C phase transition. These authors noted, first, that the smectic-C planes are perpendicular to the glass plates, and second, that the direction of the hydrodynamic movement of the fluid follows the direction of the dislocations. They designated the texture with edge dislocations as an "epi" structure. Unfortunately, this first observation of edge dislocations in the smectic-C phase has been forgotten for a long time, Meyer et al., [67] have observed edge dislocations in the smectic-C phase with a smectic-A – smectic-C phase transition in a wedge-shaped cell. Their experiment had been performed in a temperature gradient and the smectic-C has changed its characteristic tilt, which is temperature-dependent, from its maximal value to zero. This brilliant experiment has permitted to Meyer et al., [68] and to Lagerwall and Stebler [69, 70] to observe indirectly edge dislocations with a very small Burger's vector whose modulus is equal to the thickness of one smectic-C layer. Lelidis et al., [71] elaborated the study of these smectic-C dislocations in the edge-shaped cell of Cano. They proved that the edge dislocations are attached to screw dislocations. The careful comparison of the kind of the edge dislocations obtained by us with those observed by Lefèfer et al., and by Meyer et al., clearly shows that the edge dislocations observed by us are similar to those, observed by the French researchers and that they have nothing in common with the edge dislocations discovered by Meyer et al. As pointed out by Grinstein and Pelcovits [72], the wedge-shaped geometry will cause the appearance of edge dislocations in the smectic-C which is evidenced also by a geometrical point of view. On the other hand, Kléman [57] pointed out that the smectic-C defects are more complex than those in the smectic-A and often they represent confocals and dislocations, etc. Finally, one should discuss two other important questions related to such a growing of the smectic-C phase: the growth of the smectic-C phase from the cybotactic groups and the influence of the electric field on this process. The two possible orientations of the smectic-C layers in the cybotactic groups formed in the middle part of the samples (a homeotropic orientation of the preceding nematic phase) are shown in Figure 2.11.

It is quite natural to assume that the orientation of the molecules in the cybotactic groups is homeotropic due to the weak dielectric torques exerted on the smectic-A or smectic-C (one

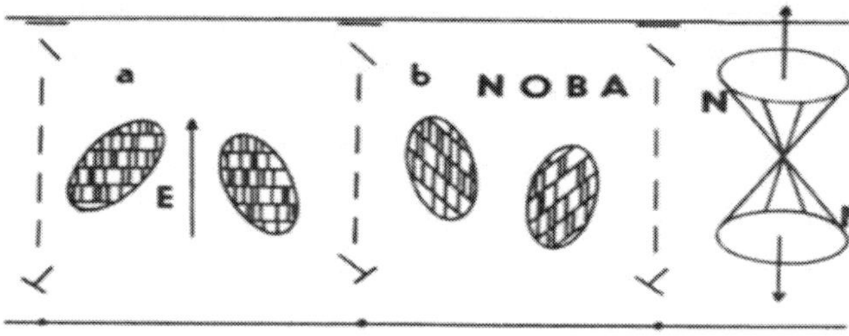

Figure 2.11. Two possible orientations of the smectic layers in the cybotactic groups being formed after rapid cooling – in the middle part of thin twisted nematic NBOA films: a) smectic-A arrangement; b) one smectic-C arrangement among the infinitely possible ones. A cone, made by all possible normals in the smectic-C layers is shown as well.

cybotactic group has no more than 100 molecules). The influence of the electric field is indirect through its action on the rest part of the nematic. As far as the smectic-A arrangement in the cybotactic groups (normal cybotactic groups) is concerned, they can easily be transformed into skewed cybotactic groups (tilt orientation of the smectic molecules with respect to the layers) through the slipping of the molecules relative to each other, a process very often observed after the smectic-A – smectic-C phase transition under the influence of various forces. Furthermore, the orientation of the molecules must be homeotropic and the orientation of the smectic-C layers must be inclined with respect to the plane of the glass plates. It is clear that the smectic-C normals make a cone with apex angle equal to 90^0 (Figure 2.11). The further formation of the smectic-C phase with the creation of edge-dislocations is schematically shown in Figure 2.12 and Figure 2.13.

Figure 2.12. One selected orientation, among the infinite possibilities of the smectic-C molecules and layers according to our model: a) initial normal position of the smectic-C molecules and b) final planar position of the smectic-C molecules which is achieved under the action of the dielectric torques on the smectic-C molecules (a negative sigh of the dielectric anisotropy of the smectic-C phase).

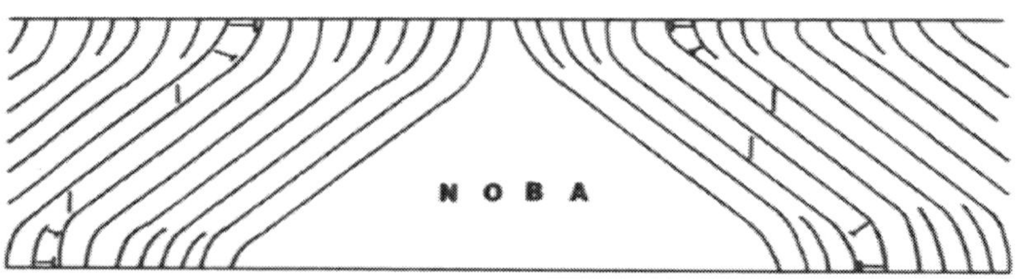

Figure 2.13. One example among infinite possibilities of a topological model showing the dislocation lines in thin smectic-C samples with oblique position of the smectic-C layers in the middle part and complex splay-twist-bend realignment of the smectic-C drector in the boundary regions.

The observed by us edge dislocations in the smectic-C were with a very large Burger's vector whose magnitude includes hundreds of molecules, participating in the formation of the dislocations. We note the following important findings characterizing these edge dislocations in the smectic-C:

1) One usually observes very sharp dark and broader bright lines adjacent to each other which do not depend significantly on the various possible positions of the nicols. They, as those in the case of nematics [73] or smectic-A [74], are due to the strong refraction and even to the internal reflection of the light which is a consequence of the great change of the optical indices of refraction of the smectic-C around the edge dislocation effected by so-called Becke lines [73,74].
2) The dislocation lines, parallel to the smectic-C planes, are localized near the glass plates which results in their relief being very clear and distinct from the rest part of the smectic-C layer.
3) The inclination of the smectic-C layers in the middle part of the films, the twisting of the director in the nematic phase and the ability of the liquid crystal under study to form cybotactic groups can easily lead to a significant bending of the smectic-C layer and to the creation of dislocation lines. This bending of the smectic-C layers is accompanied by complex bend-twist-splay deformations of the director of the smectic-C phase (see Figure 2.13). It is confirmed by the strong influence of the optical properties of these dislocations on the electric field amplitude even in the smectic-C phase (we note here that the dielectric anisotropy of the smectic-C phase of NOBA is in the range of -0.025) together with the low voltage of 80-300 $V_{r.m.s.}$ being applied across the cells.
4) The increase of the voltage up to 300 $V_{r.m.s.}$, i.e., 10 times the threshold voltage for the nematic phase, leads not only to a decrease of the dislocation period, but also to a decrease of the penetration depth of the dislocation walls into the middle part of the films.
5) Edge dislocation lines cannot be obtained after the rapid cooling of twisted nematc films down to the smectic-C phase when the smectic-C layers are parallel to the glass plates due to the easy adaption of the twist deformation.
6) Edge dislocations in the smectic-C phase cannot be formed after cooling of the splay-bend realigned nematic in the boundary regions and only local smectic-C monocrystals with oblique position of the smectic-C layers have been formed.
7) Cooling of very thin twisted nematic layers with a thickness of 2 μm led to the formation of π- inversion walls, smectic-C local monocrystals and edge dislocations, very similar to those, already observed by Meyer et al., [67, 68].
8) The strong boundary coupling of the nematic and smectic-C directors and especially strong azimuthal coupling of the nematic-C director are very important for the formation of edge dislocations in the smectic-C.
9) The nematic-smectic-C phase transition should be of a second order, or nearly of a first order.
10) The relaxation of the smectic-C edge dislocations in less energetic disclination lines or focal-conics was not observed during the performance of our experiment.

2.9. Discussion

As noted in the beginning of this chapter, relations and inequalities have been obtained between the elastic constants of the smectic-C based on stability conditions (2.4), [2, 3, 75]. All the elastic constants of the smectic-C, with only one exception of the modulus $\overline{B}$, are very similar to the elastic constants of the nematics, and are expected to be in the range of 10^{-6} dyn. The measurement of the elastic constants A_{11}, A_{12} and A_{21}, related to electric or magnetic fields cannot ensure the eventual determination of these constants due to the very weak, usually non-observed deformations in the smectic-C phase. Another useful way for determining these constants is connected with measurement of the periodicity of the smectic-C layer undulations, investigated for the first time by Johnson and Saupe. However, these measurements cannot give the difference between the elastic constants $A_{11}$and A_{21}, and A_{11} and A_{12}. The paper of Allet et al., for the transformation of edge disclination with a strength of +½ and a partial twist into the same disclination without twist under the application of a magnetic field with above-threshold value [77], (see also the References [76] and [78]) and its elaboration (in a magnetic field) [79-81] permits the evaluation of some of the elastic constants A as follows:

a)

$$H_c^2 = (2A_{21} - A_{11})\frac{\ln(R/r_0)}{\Delta\chi \sin^2\psi(R^2 - r_0^2)}$$

when $A_{11} < 0$ and $2A_{21} - A_{11} > 0$ (2.15)

b)

$$H_c^2 = (2A_{12} - A_{11})\frac{\ln(R/r_0)}{\Delta\chi \sin^2\psi(R^2 - r_0^2)}$$

when $A_{11} < 0$ and $2A_{12} - A_{11} > 0$ (2.16)

where R is the radius of the semi-cylinder showing the + ½ edge smectic-C disclination with a partial twist, r_0 is the core radius of the disclination, ψ is the angle of inclination of the smectic-C molecules (the magnetic field is applied along the semi-axis of the cylinder).

c)

$$H_c^2 = \frac{(A_{11} - 2A_{21})\ln(R/r_0)}{\Delta\chi(R^2 - r_0^2)\sin^2\psi}$$ when $A_{11} > 0$, $A_{11} - A_{21} > 0$ (2.17)

d)

$$H_c^2 = \frac{(A_{11} - 2A_{12})\ln(R/r_0)}{\Delta\chi(R^2 - r_0^2)\sin^2\psi}$$ when $A_{11} > 0$, $A_{11} - A_{12} > 0$ (2.18)

(the magnetic field is applied in the perpendicular direction relative to cases (a) and (b)).

It is natural, the magnetic field to be replaced by AC electric field with sufficiently frequency to avoid the electro-hydrodynamic effects for the case of smectic-C liquid crystals with positive dielectric anisotropy (the required strength of the applied magnetic field should be in the range of several thousand gauss [79], an usefull transformation from the magnetic units in electrical ones can be found, for instance, in the book by de Gennes [82]).

Finally, let us stress that the growth of the smectic-A and smectic-C from a strongly deformed in the boundary regions nematic is quite different. The first difference in the growth of these two smectics is that the smectic-A is sensitive to the deformations of bend and twist in the nematic while the smectic-C is sensitive only to twist. The second difference is that the smectic-C dislocations are parallel to the smectic-C layers which are perpendicular to the glass plates of the liquid crystal cells. The third difference is that the dislocations in the smectic-A usually relax in a system of confocal domains, which has not been observed in the smectic-C although it is evident that the elastic energy of the smectic-C confocals is smaller than that of the edge dislocations also in the smectics-C. According to the theoretical results of Nakagawa [61], the same elastic constants of the smectic-C must vanish when in the structure of the smectic-C toric domains there are not disclinations:

$$B_{13} = C_1 = C_2 = 0 \tag{2.19}$$

This relation for the case of the smectic-C of NOBA is apparently satisfied [55], and leads to the conclusion that three elastic constants A_{11}, A_{12} and A_{21} must be positive which means that A_{11} is positive.

CONCLUSION

1) Originally, a smectic-C liquid crystal with a nematic – smectic-C phase transition of a second order or a weak first order is formed from the middle oriented part and deformed part in the boundary regions of a preliminary 90^0-twisted nematic excited by a high-frequency electric field. At these conditions, one obtains either a Schlieren texture with planar smectic-C layers and Fergason-Johnson-Saupe stripes (the applied voltage is between 1.5 U_c and $3U_c$, where U_c is the Fréedericksz threshold voltage) or local smectic-C monocrystals with edge dislocations and a vertical position of the smectic-C layers when the voltage is bigger.
2) It is confirmed qualitatively that the smectic-C phase grows at a lower phase transitional temperature from a strongly deformed nematic phase with deformations of splay-bend-twist.
3) A new method is developed for the formation of optically observable edge dislocations in planar smectic-C cells.
4) The formation of Fergason-Johnson-Saupe stripes can be easily realized when the smectic-C phase is formed after fast cooling of high-tilted nematic with weak surface anchoring.
5) Estimation is made of some of the elastic constants and of the stretching modulus of the layers of the smectic-C under study.

6) A new method for measurement of the elastic constants A of the smectic-C is suggested on the basis of the behavior of a disclination with a strength of s = + ½ and a partial twist under the action of magnetic or high-frequency electric fields which with a jump is transformed into a disclination s = + ½ without twist.
7) All possible methods for measurement of the elastic constants A or of their combinations are discussed.

REFERENCES

[1] Luckhust, G. R.; Sundholm, F. *Mol. Phys.* 1971, 21, 349-352.
[2] Saupe, A. *Mol. Cryst. Liq. Cryst.* 1969, 7, 59-74.
[3] Orsay Group on Liquid crystals *Sol. St. Commun.* 1971, 9, 653-655.
[4] Rapini, A. *J. Phys.* (Paris) 1972, 33, 237-247.
[5] Leslie, F. M.; Stewart, I. W.; Nakagawa, M. *Mol. Cryst. Liq. Cryst.* 1991, 198, 443-454.
[6] Leslie, F. M.; Stewart, I. W.; Carlsson, T.; Nakagawa, M. *Cont. Mech. Thermodyn.* 1991, 3, 237-250.
[7] Atkin, R. J.; Stewart, I. W. *Liq. Cryst.* (UK) 1997, 22. 585-594.
[8] Carlsson, T.; Stewart, I. W.; Leslie, F. M. *Liq. Cryst.* (UK) 1991, 9, 661-678.
[9] Carlsson, T.; Stewart, I. W.; Leslie, F. M. *Liq. Cryst.* (UK) 1992, 11, 49-61.
[10] Kedney, P. J.; Stewart, I. W. Z. angew. *Math. Phys.*1994, 45, 882-888.
[11] Atkin, R. J.; Stewart, I. W. Q. *Jl. Mech. Appl. Math.* 1994, 47, 231-245.
[12] Atkin, R. J.; Stewart, I. W. Eur. *J. Appl. Math.* 1997, 8, 253-262.
[13] Barratt, P. J.; Duffy, B. R. *Liq. Cryst.(*UK) 1997, 23, 525-529.
[14] Stewart, I. W. *Liq. Cryst.(*UK) 2003, 30, 909-920.
[15] Meirovitch, E.; Luz, Z.; Alexander, S. *Phys. Rev.* A 1977, 15, 406-416.
[16] Pelzl, G.; Schubert, H.; Zaschke, H.; Demus, D. *Kristall und Technik* 1979, 14, 817-823.
[17] Pelzl, G.; Kolbe, P.; Preukschas, U.; Diele, S.; Demus, D. *Mol. Cryst. Liq. Cryst.* 1979, 53, 167-180.
[18] Schiller, P.; Pelzl. G. *Cryst. Res. Technol.* 1983, 18, 923-931.
[19] Galerne, Y.; Martinand, J. L.; Durand, G.; Veyssie, M. *Phys. Rev. Lett.* 1972, 29, 562-564.
[20] Pelzl, G.; Schiller, P.; Demus, D. *Liq. Cryst.* (UK) 1987, 2, 131-148.
[21] Dierking, I.; Giesselmann, F.; Zugenmeier, P.; Pelzl, G.; Schiller, P *Cryst. Res. Technol.* 1992, 27, 727-739.
[22] Brown, C. V.; Jones, J. C.; Dunn, P. E. *Mol. Cryst. Liq. Cryst.* 1999, 331, 465-472.
[23] Brown, C.V.; Dunn, P. E.; Jones, J. C. *Eur. J. Appl. Math.* 1997, 8, 281-291.
[24] Lejček, L. Czech. J. Phys. B 1984, 34, 563-570.
[25] De Gennes, P. G. *Mol. Cryst. Liq. Cryst.* 1973, 21, 49-76.
[26] Chen, J.-H.; Lubensky, T. C. *Phys. Rev. A* 1976, 14, 1202-1207.
[27] Chu, K. C.; McMillan, W. L. *Phys. Rev. A* 1977, 15, 1181-1187.
[28] Swift, J. P*hys. Rev. A* 1976, 14, 2274-2277.
[29] Leitner, P.; Swift, J. *Phys. Rev. A 1979*, 19, 894-898.
[30] Witanachi, S.; Huang, J.; Ho, J. T. *Phys. Rev. Lett.* 1983, 50, 594-597.

[31] De Vries, A. J. *Phys. Colloq.* (Paris) 1975, 36, C1-1 – C1-11,
[32] Azaroff, L. V.; Schuman, C. A. *Mol. Cryst. Liq. Cryst.* 1985, 122, 309-319.
[33] De Vries, A. J. *Mol. Liq.* 1986, 31, 193-202.
[34] Herbert, A. J. *Trans. Far. Soc.* 1967, 63, 555-560.
[35] Rao, N. V. S.; Pisipati, V. G. K. M. *Z. Naturforsch.* (a) 1985, 40, 466-468.
[36] Taylor, T. R.; Fergason, J. L.; Arora, S. L. *Phys. Rev. Lett.* 1970, 24, 359-362.
[37] Chou, L. S.; Carr, E. F. In Liquid crystals & Ordered Fluids; Johnson, J. F.; Porter, R. S.; Eds.; *Plenum Press:* New York, U.S. 1973; vol. 2, pp 39-51.
[38] Chou, L. S.; Carr, E. F. *Phys. Rev. A* 1973, 7, 1639-1641.
[39] Carr, E. F. *Phys. Rev. A* 1975, 12, 327-329.
[40] Kuvatov, Z. H.; Kapustin, A. P.; Trofimov, A. N. Pisma *Zh. Eksp. Teor. Fiz* (USSR) 1974, 19, 89-92.
[41] Kuvatov, Z. H.; Kapustin, A. P.; Trofimov, A. N. In Liquid crystals and Applications, Collection of the University, Ivanovo (USSR) 1976, 74-85.
[42] Kuvatov, Z. H.; Kapustin, A. P.; Trofimov, A. N. *Kristallografiya* (USSR) 1979, 24, 397-400.
[43] Demus, D.; Richter, L. Textures of Liquid crystals; *VEB Deutcher Verlag,* DDR; 1978.
[44] Arora, S. L.; Fergason, J. L.; Saupe, A. *Mol. Cryst. Liq. Cryst.* 1970, 10, 243-257.
[45] Fergason, J. L. *Private communication* 1970.
[46] Johnson, D.; Saupe, A. *Phys. Rev. A* 1977, 15, 2079-2085.
[47] Bartolino, R.; Durand, G. *Ann. Phys.* (France) 1978, 3, 257-262.
[48] Kumar, S. *Phys. Rev. A* 1981, 23, 3207-3214.
[49] Fournier, J. B.; Galatola, P. *J. Phys. II France* 1995, 5, 1297-1320.
[50] Torgova, S. I.; Sparavigna, A.; Strigazzi, A. *Mol. Cryst. Liq. Cryst.* 2000, 352, 111-118.
[51] Price, F. P.; Bak, Ch. S. *Mol. Cryst. Liq. Cryst.* 1975, 29, 2250229.
[52] Schiller, P. *Ferroelectrics* 1991, 122, 261-277.
[53] Giocondo, M. Jákli, A.; Saupe, A. *Eur. Phys. J. E* 2000, 1, 61-65.
[54] Durand, G. C.R. *Acad. Sc Paris* B1972, 275, 629-631.
[55] Hinov, H. P.; Petrov, M. *Mol. Cryst. Liq. Cryst.* 1983, 100, 223-251.
[56] Petrov, M. P.; Simova, P. D. *Phys. Sat. Sol.* (a) 1986, 98, 399-407.
[57] Kléman, M. Points, Liqnes, Parois dans les Fluides Anisotropies et les Solides cristallins Les Edition de Physique, Orsay, France; 1977; t. II. pp 41-42.
[58] Findon, A.; Gleeson, H. F.; Lydon, J. E. *Liq. Cryst.*(UK) 2002, 29, 79-85.
[59] Perez, A.; Brunet, M.; Parodi, O. J. *Phys. Lett.* (Paris) 1978, 39, L-353 – L-357.
[60] Bourdon, L.; Sommeria, J.; Kléman, M. *J. Phys.* (Paris) 1982, 43, 77-96.
[61] Nakagawa M. J. *Phys. Soc. Jpn.* 1990, 59, 81-89.
[62] Collin, D.; Gallani, J. L.; Martinoty, P. *Phys. Rev. Lett.* 1987, 58, 254-257,
[63] Walker, A. J.; Stewart, T. W. *J. Phys. A: Math. Theor.* 2007, 40, 11849-11861.
[64] Stewart, I. W. The Static and Dynamic Continuum Theory of Liquid crystals; *Taylor and Francis,* London, 2004.
[65] Wulf, A. *Mol. Cryst. Liq. Cryst.* 1978, 47, 225-236.
[66] Lefèvre, M.; Martinand, J.-L.; Durand, G.; Veyssié, M. *C.R. Acad. Sc.* Paris B1971, 273, 403-406.
[67] Meyer, R. B.; Stebler, B.; Lagerwall, S. T. *Phys. Rev. Lett.* 1978, 41, 1393-1395.
[68] Lagerwall, S. T.; Meyer, R. B.; Stebler, B. *Ann. Phys.* (France) 1978, 3, 249-256.
[69] Lagerwall, S. T.; Stebler, B. *J. Phys. Colloq.* (Paris) 1979, 40, C3-53 – C3-57.

[70] Lagerwall, S. T.; Stebler, B. *J. Phys. Bull.* 1980, 31, 349-350.
[71] Lelidis, I.; Kléman, M.; Martin, J. L. *Mol. Cryst. Liq. Cryst.* 2000, 351, 187-196.
[72] Grinstein, G.; Pelcovits, R. A. *Phys. Rev.* A 1982, 26, 2196-2217.
[73] Hinov, H. P.; Vistin', L. K.; Marinov, Y. G. *J. Phys. Chem.* 2014, 118, 4220-4227.
[74] Williams, D. C. Dèfauts de Structure dans les Smectique A, Thése Universitè de Paris-Sud, *Centre d'Orsay,* France, 1976, p 30.
[75] Kléman, M.; Lejček, L. *Phil. Mag.* A 1980, 42, 671-682.
[76] Scheffer, T. J.; Grüler, H.; Meier, G. *Sol. St. Commun.* 1972, 11, 253-257.
[77] Allet, C.; Kléman, M.; Vidal, P. *J. Phys.* (Paris) 1978, 39, 181-188.
[78] Bouligand, Y.; Kléman, M. *J. Phys.*(Paris) 1979, 40, 79-97.
[79] Hinov, H. P. *Compt. Rend. Acad. Bulg. Sci.* 1983, 36, 611-614.
[80] Hinov, H. P. *J. Phys. Lett.* (Paris) 1984, 45, L-185 – L-191.
[81] Hinov, H. P. *Fifth LC Conf. Soc. Countries, Proc.* October 10-15 1983, 187-189.
[82] De Gennes, P. G. The Physics of Liquid crystals, *Clarendon Press,* Oxford, England, 1974, chapter 3.

Chapter 3

ELECTRO-OPTIC BEHAVIOR AND TEXTURES IN LARGE-PITCH CHOLESTERICS

3.1. GENERAL REMARKS

It is well known that formally the cholesterics are twisted nematics. However, there is one essential difference. The liquid crystal molecules of the cholesterics are chiral and consequently their minimal elastic energy in the ground state is not vanishing as in the case of the nematics and has a definite value due to the pitch of twisting p = π/d, where d is the distance along which twisting of the cholesteric director is with π. Due to these properties, the cholesterics are layered structures although they do not have a change in their density as in the case of the smectics. The cholesterics are usually divided into large-pitch cholesterics when the pitch of twist is equal, or bigger, relative to the thickness of the liquid crystal cells under study, i.e., when the ratio d/p < 1 and to short-pitch cholesterics when the ratio d/p > 1. As it is well-known, the pitch of the conventional cholesterics is of the order of several micrometers, and even, of parts of the micrometer. Evidently, such cholesterics are optically active. In our experiments, the short-pitch cholesterics were normally used as a dopant to the nematics in order to create large-pitch cholesterics. Our aim was to study the influence of the strength of the surface anchoring and the value (qualitatively) of the surface tilt on the large-pitch cholesteric textures as well as their behavior under the influence of externally applied high-frequency electric field changing either by steps or linearly. At the latter case, the dynamic hysteresis was studied depending on the surface anchoring and the initial textures of the large-pitch cholesteric. The large-pitch cholesterics show several basic textures: confocal with cholesteric pitch in the plane of the Electrodes (glass plates) and a complicated texture with spirals of either handednesses, a texture called „finger prints" without spirals which is usually embedded in a homeotropic nematic and Grandjean orientation (or texture) when the axis of twist is along Z, perpendicular to the plane of the glass plates. The cholesteric-nematic-cholesteric transition which is of the type of the Fréedericksz transition in nematics is very complicated. For example, the application of an electric field perpendicular to the initial texture „finger prints", i.e., perpendicular to the glass plates (positive dielectric anisotropy of the cholesteric) leads to a threshold unwinding of the cholesteric, in accordance with a relation obtained by de Gennes [1] and Meyer [2] which has the following form:

$$E_{thresholdC-N} = \frac{\pi^2}{p_0}\left(\frac{4\pi K_{22}}{\Delta\varepsilon}\right)^{1/2} \tag{3.1}$$

In the formula (3.1) by p_0 the natural pitch of the cholesteric is designated.

The exact formula, however, was obtained by Leslie [3]. Later, all these formulae have been confirmed in detailed theoretical calculations by Kedney and Stewart [4], Regaya et al., [5], Dreher [6], Malkin [7], Gotra et al., [8]. The experimental study of Wysocki et al., [9], Meyer [10] and Durand et al., [11] have completed the study for the dechiralizing of the cholesterics under external electric or magnetic fields. The hysteresis in the relaxation of the pseudonematic to the initial Grandjean texture or to the intermediate confocal texture has been observed and studied by many authors [12-15]. At the relaxation of the pseudohomeotropic nematic to the Grandjean orientation, a hysteresis appears and the liquid crystal relaxes to the confocal texture $(p_0K_{33}/d) < 2K_{22}q)$ according to the formula obtained by Greubel [16]

$$E_{threshold} = \left(\frac{\pi}{p_0}\right)\left(\frac{4K_{22}^2 - \left(\frac{p_0 K_{22}}{d}\right)^2}{|\Delta\varepsilon| K_{33}}\right)^{1/2} \tag{3.2}$$

where $|\Delta\varepsilon|$ is the modulus of the dielectric anisotropy, d is the thickness of the cholesteric layer, K_{22} and K_{33} are the elastic constants of twist and bend and p_0 is the natural pitch. The comparison of the above two formulae shows that there is a hysteresis and the pseudohomeotropic nematic will relax to the confocal cholesteric texture when the following inequality $(p_0K_{33}/d) < 3K_{22}$ is fulfilled.

However, the pseudo-homeotropic nematic can transform at certain value of the voltage to a new Grandjean texture determined from the relation:

$$E_{HP^*} = (2/\pi)(K_{22}/K_{33})^{1/2} E_c \sim 0{,}42 E_c \tag{3.3}$$

(this effect has been studied in a number of papers [12, 14, 15, 17-19]). The pitch p* is greater relative to the natural pitch of the cholesteric p: $p^* \sim p_0K_{33}/K_{22}$. It is clear that the cholesteric with such a Grandjean texture can be transformed into usual Grandjean texture via two-dimensional instabilities of Helfrich-Hurault [21-25] under the action of the electric field. The relaxation to the Grandjean orientation with a bigger cholesteric pitch, however, is very often local and very fast with reaction rates of the order of 1-2 ms and, consequently, the resulting texture is strongly scattering. It is clear that at these conditions, the relaxation to the initial Grandjean texture is very difficult. The unwinding of the cholesteric is much more difficult when the cholesteric is with Grandjean orientation (positive dielectric anisotropy) and the high-frequency electric field is applied along the axis Z. At these conditions, the deformation of the spiral leads to its tilt and to a formation of various textures [26]. The "turning" of the twist direction from Z to the plane (X, Y) is performed by the creation of confocal texture. This deformation is often called "conical". Leslie [3] has calculated the threshold:

$$E_c = \left(\frac{4\pi}{|\Delta\varepsilon|}\right)^{1/2}\left[K_{11}\left(\frac{\pi}{d}\right)^2 + K_{33}\left(\frac{2\pi}{p_0}\right)^2\right]^{1/2} \tag{3.4}$$

It is clear that „turning" of the spiral is effected by the transition of de Gennes-Meyer. One should take into account additionally the kind of cholesteric orientation at the surfaces: planar, tilted or homeotropic; the strength of the surface anchoring: weak, when the interfacial surface energy is of the order of 10^{-4} erg/cm^2, middle when it is of the order of 10^{-2} erg/cm^2 and strong when it is bigger than 0.1 erg/cm^2 and finally, one should take into account the surface tilt of the cholesteric director. The simplest boundary orientation of the large-pitch cholesteric is the homeotropic one. At such a boundary orientation of the cholesteric director, the "finger-print" is the preferred large-pitch cholesteric texture with the lowest elastic energy when d ~p. Models have been proposed without the existence of defects [27, 28] as well as models which include defects in the "finger-print cholesteric texture " [29]. According to Stieb [28], the conical deformation replaces the pure twist at a critical thickness, determined by

$$d_c = (1/2)p_0(K_{33}/K_{22}) \tag{3.5}$$

This formula corresponds to the transformation of one pseudo-homeotropic nematic towards the Grandjean orientation with a bigger pitch relative to the natural pitch. In the opinion of Stieb [28], the homogeneous conical deformation is not stable and relaxes slowly towards "finger-print". According to Stieb, in the "finger-print" texture there are now disclinations when the pitch of the cholesteric is in the range of the thickness of the liquid crystal layer and the surface anchoring of the cholesteric is weak. The model of Cladis-Kléman [29] is valid when d/p >> 1 and the surface anchoring of the cholesteric is not important since the cholesteric texture is far away from the boundaries. Dislocations also can be observed in the cholesterics since they are with a layered structure [30] and sometimes they interact with the boundaries [31]. In our opinion, however, the disclinations are of great importance for the dynamics of the cholesteric-nematic-cholesteric transition and are connected with bistability [32, 33, 34]. The instability of Helfrich-Hurault is with the lowest threshold voltage among all known instabilities in cholesterics. For the case of strongly anchored cholesterics, the following formulae describing this instability are known:

$$U_c^2 = \frac{8\pi^3}{|\Delta\varepsilon|}\left(6K_{22}K_{33}\right)^{1/2}\left(d/p_0\right) \tag{3.6}$$

$$\lambda_c^2 = \left(3K_{33}/2K_{22}\right)^{1/2} dp_0 \tag{3.7}$$

Bartolino et al., [35] proved experimentally that the instability of Helfrich-Hurault is the lowest voltage one, and after that the spiral is "turned to $\pi/2$" and one appears the confocal texture. Earlier, Scheffer [36] has observed grid-instability when d >> p_0 (the grid texture is a superposition of two one-dimensional instabilities developed in two mutually orthogonal directions). On the other hand, Hervet et al., [37] have observed one-dimensional deformation of large-pitch cholesteric when d~ p_0 and a two-dimensional grid when d >> p_0. Some authors noted that the unwinding of the cholesteric should be performed by steps when the cholesteric is strongly-anchored at the two boundaries. The instability of Helfrich-Hurault can also be observed at the dynamic process of the twisting or untwisted cholesterics [38]. Ishikawa and Lavrentovich [39] have also considered the instability of Helfrich-Hurault at a finite surface energy of the cholesteric director.

All textures and dynamic electro-optic curves, obtained at switch-on and switch-off of the electric field as well as all hysteretic curves were obtained during the study of different anchored cholesteric layers with small or large tilt relative to the glass plates normal. The relatively strong anchoring was achieved by rubbing of the glass plates with abrasive paste with grains between 0.5 and 2 μm. The weak anchoring was achieved by using of a surfactant – usual soap. The surface tilt at both glass plates depends on the mutual position of the glass plates relative to each other when they are rubbed in one and the same direction. In this way, we have obtained nearly planar oriented or highly tilted oriented liquid crystal layers with strong-strong or weak-weak anchoring. Asymmetrical strong-weak surface anchoring was achieved when one of the two glass plates was treated to obtain a strong surface anchoring and the other glass plate was treated to obtain a weak anchoring. We have used in our experiments a liquid crystal mixture consisting of 89 wt. % MBBA [40-42] (a liquid crystal which is strongly influenced by the soap) – 10 wt. % 5CB [42-47] (a liquid crystal which is weakly influenced by the soap) – 1wt. % CC (cholesteryl chloride) (the dielectric anisotropy of this mixture is positive of values between 1.5 and 2). We present below the chemical formulae of MBBA, 5CB and CC, and their phase transitions:

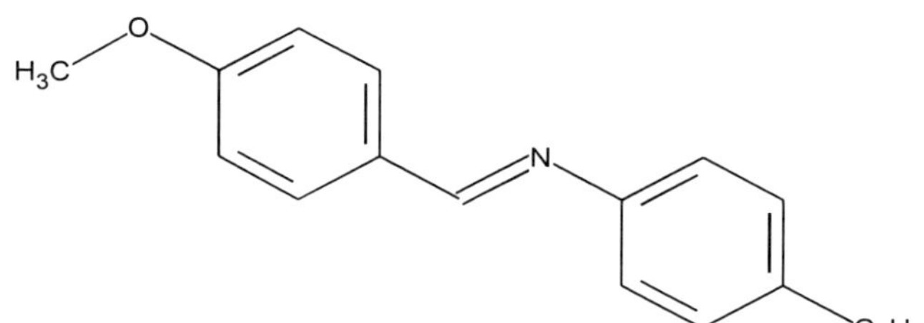

MBBA: I 48^0C N 22^0C C

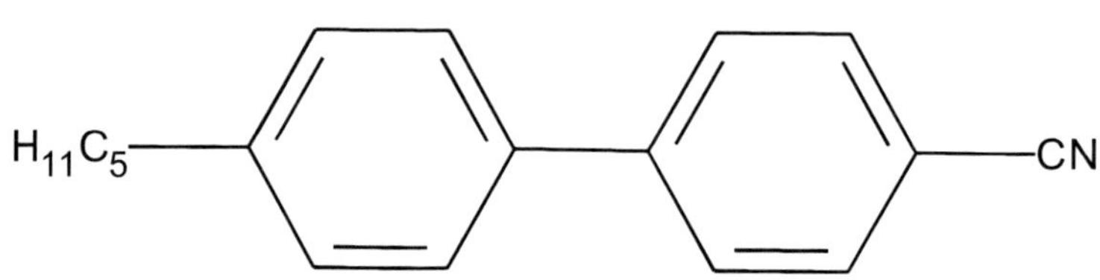

5CB: I 37^0C N 23^0 C C

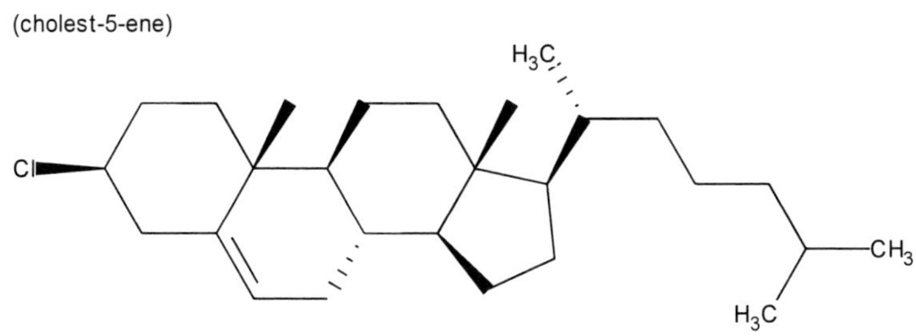

CC: I (62^0C Ch) 97^0C C

As far as the surface anchoring of the cholesterics is concerned, we should stress that the boundary interactions are different for the cases of nematics and cholesterics. For example, the surface anchoring for the case of nematic can be relatively strong (the surface forces can penetrate into 150-200 μm) whereas the same boundary can influence weak by the short-pitch cholesterics (the surface forces can penetrate into considerably smaller distance of 50 μm)[48]. We have obtained finger-print textures embedded into a homeotropic nematic with a relatively weak surface anchoring treating some of the glass plates of the liquid crystal cells only with abrasive paste containing small metal spheres. The use of soap, however, led to considerable decrease of the surface coupling and to a difference in the electro-optic curves. We should stress here that the large variety of the textures and of the electro-optic curves is due to the small dielectric anisotropy $\Delta\varepsilon \sim 1.5 - 2$. At larger values of $\Delta\varepsilon$ and when the ratio $d/p \sim 0.2$ which is obtained in our experiment and at a weak surface anchoring, one should realize only a Fréedericksz transition expressed after de Gennes [40] by the following relation:

$$V_{FT} = \left(\frac{K_{11} 4\pi^2}{|\Delta\varepsilon|} + \frac{16\pi^3 d^2 K_{33}}{|\Delta\varepsilon| p_0^2} \right)^{1/2} \tag{3.8}$$

Sartirana et al., [49] obtained a second relation of the threshold voltage leading to tilt of the axis of twisting and finally, to its unwinding:

$$V_{tilt} = 2\pi \left(\frac{4K_{22}\pi}{|\Delta\varepsilon|} \right)^{1/2} d / p_0 \tag{3.9}$$

The value of this threshold is lower relative to the well-known threshold obtained by de Gennes and Meyer (see the relation (3.1). The weak surface coupling of the large-pitch cholesteric with the bonding glass plates permitted to investigate its electro-optics and hysteretic behavior for the case of d/p) ~ 0.2 (we should stress here that for the case of a relatively strong anchoring and the same conditions, usually one observes a Fréedericksz transition and the unwinding of the large-pitch cholesterics under the influence of the surface forces [50-52]). Meyer is the first author [10] who noticed that the surface forces can lead to hysteresis at the process of unwinding of the cholesteric under the influence of external magnetic field, applied perpendicular to the twist axis. Greubel [16], Gerber [53] and van

Sprang and van de Venne [54] have considered the voltage leading to the unwinding of the cholesteric by cholesteric-nematic transition for homeotropic and planar orientation of the cholesteric and a finite surface energy in the range of 10^{-2} erg/cm^2 and have obtained the following approximate formula:

$$V_{threshold} = \left[\frac{8\pi^2 d K_{22}}{p_0^2 |\Delta\varepsilon|} - \frac{8F_{\Pi} d}{|\Delta\varepsilon|} \right]^{1/2} \tag{3.10}$$

In this formula F_{Π} designates the surface energy on unit area. This empiric formula has been mentioned in the review by Chilaya [26] and also by other authors [55, 56]. It is well-known that inhomogeneous deformations can exist during the unwinding process of the cholesteric under external electric or magnetic fields, applied normally to the twisting of the cholesteric [57], Goossens [58] is the first author who considered theoretically the problem and obtained approximate solution for the value of the threshold electric field both for homeotropic or planar orientation of the cholesteric at the boundaries:

$$E_{CH}^2 = \frac{K_{22}}{|\Delta\varepsilon|} \left[\left(\frac{q_0 \pi}{2} \right)^2 \mp 2q^2 \cosh^{-2}\left(\frac{\gamma q d}{2} \right) \right]$$

$$\gamma^2 = K_{22} / K_{33} \text{ , } q_0 = 2\pi / p_0 \tag{3.11}$$

here - and + correspond to the case of homeotropic, resp. planar orientation, and q is determined by the following formula:

$$q = q_0 \left(1 - \frac{1 + 2\ln 2qa}{4\gamma q_0 d} \right) \tag{3.12}$$

The second term in eq. (3.11) represents the deformation of the director in the "finger-print" or confocal textures for the case of a fixed boundary orientation of the cholesteric director, This term has a small value when d/p > 2. The theory of Goossens has been elaborated in Reference [56]. The influence of the boundary conditions has been considered by many authors [59-61]. Mochizuki and Kobayashi [60] have obtained a formula for unwinding electric field in the case of thin cholesteric films:

$$E_{CN}^* = \sqrt{E_{CN}^2 - \frac{2A_0}{|\Delta\varepsilon|(d - \varsigma)}} \tag{3.13}$$

In this formula A_0 is the θ-surface energy, d is the thickness of the liquid crystal layer and ς is the thickness of the winded layer in the moment of the application of the electric field.

During the performance of our experiments we have observed all known and typical for the large-pitch cholesterics textures such as Grandjean-like texture with existence of "oily streaks" (the various textures of the cholesterics can be found in References [62-64]), confocals with twisting in the plane of the glass plates which serve as Electrodes and spirals with both handednesses, various textures of "finger-print", embedded into a pseudo-homeotropic nematic [65]. At switch-on or switch-off of a big voltage of 10-15 $V_{r.m.s.}$ one can observe complex system which is called "disclination patterns" when the applied voltage is high-frequency[65] or a scroll texture at the application of a DC voltage (Figure 3.1).

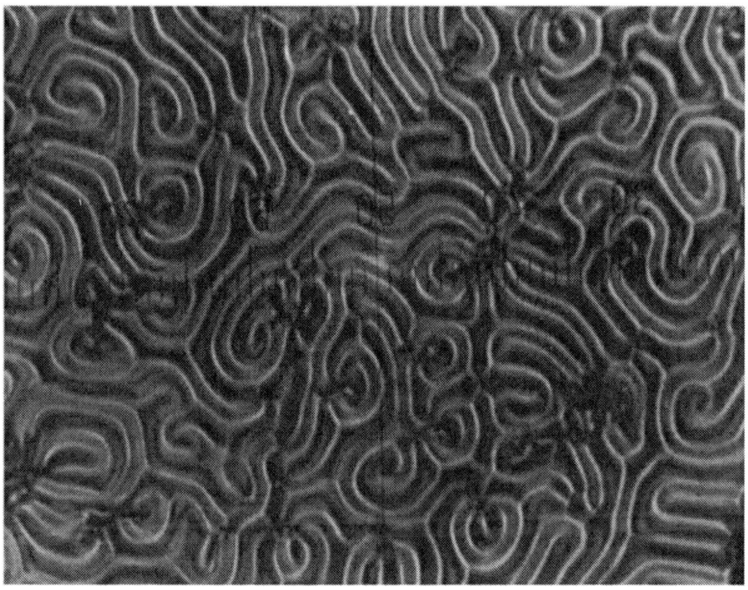

Figure 3.1. "Scroll" texture obtained in a MBBA-5CB-CC mixture with a thickness of 15 μm and hybrid orientation after applying and removing of a DC voltage of 10–15V. 10 small divisions correspond to 45 μm; crossed nicols, room temperature.

Both textures form near the glass plates at a conflict between the surface and bulk orientation of the cholesteric [32, 52, 66].

3.2. Strong-Strong, Strong-Weak and Weak-Weak Anchored Reverse-Tilted Large-Pitch Cholesterics with the Mixture MBBA – 5CB – CC

The use of soap in a dry state and its rubbing on the already-rubbed substrate is schematically shown in Figure 3.2.

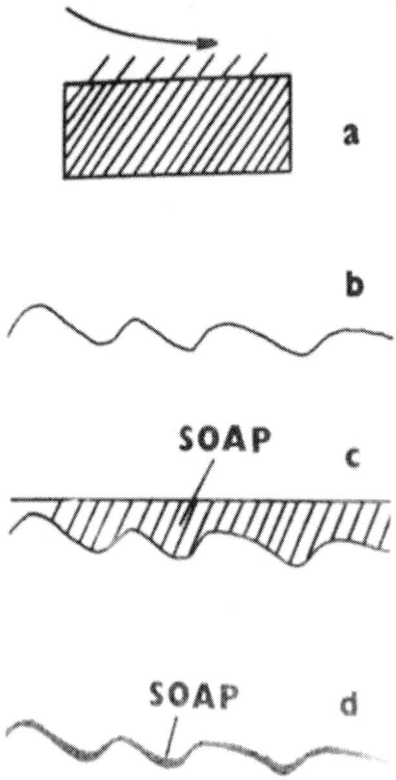

Figure 3.2. A way of the glass plates treatment:a) the rubbing direction can predetermine the sign of the director's surface tilt;b) the kind of the grooves after the first rubbing;c) covering of the glass plates with a thick soap layer;d) the grooves with part of the soap after the second rubbing with a clear cloth.

Of special importance is the experimental finding that soap-treated glass plates determine weak θ-polar non-degenerated surface anchoring (the angle θ determines the so-called "tilt" at the boundaries). The sign of θ at the boundaries can be "pre-determined" by the way of the soap deposition in a dry state Moreover, the θ-tilt depends strongly on the thickness of the soap deposition. The surface tilt of the director decreases to $30\text{-}40^0$ with respect to the tangent to the glass plates for thin soap layers with a thickness of parts of a micron. There is no doubt that the liquid crystal penetrates in many pores existing in the soap layer and as a consequence the interaction between the liquid crystal and the soap is much stronger. Furthermore, the soap can considerably reduce the amount of the surface irregularities and at the same time can influence stability of the cholesteric textures since the transition of an oriented pseudo-nematic in confocal or other type texture starts from the surface defects. The strong-weak anchoring of the large-pitch cholesterics is rarely considered in the literature [67, 68]. We have determined such an asymmetrical anchoring when one of the two glass plates confining the liquid crystal was treated by soap. On the other hand, we have determined weak-weak anchoring of the cholesteric layer when both glass plates were treated by soap (Figure 3.3).

The two glass plates can be mutually oriented in various ways but we chose two ways which give either highly-tilted cholesteric layer with a weak surface coupling or nearly planar layer with a weak surface coupling and a reverse tilt at both boundaries (see Figure 3.3) Here we should mention that it was important to obtain planar layers without a soap treatment of the glass plates due to the ability of the 5CB molecules to orient themselves homeotropic to untreated glass plates. Let us now consider separately the important cases of differently oriented layers of the mixture MBBA-5CB-CC [69].

a. Reversely-tilted Strong-Strong Anchored MBBA-5CB-CC Cholesteric Layers

Reversely-tilted cholesteric layers with a large pitch ($d/p_0 < 0.25$) have been obtained by usual rubbing of the glass plates and have shown "finger-print" textures embedded into a homeotropic matrix. This matrix is very stable and reproducible which shows that the director at the boundaries is highly tilted relatively to the glass plate tangent.

b. Reversely-tilted Strong-Weak Anchored MBBA-5CB-CC Cholesteric Layers

The strong-weak anchored Reversely-tilted cholesteric layers, display a very large variety of textures, depending on the soap thickness. For example, we have obtained either many Grandjean-like regions separated by walls and disclinations when a thick soap layer is used. These regions increase in their size when the thickness of the soap decreases (see Figure 3.2, case d) also divided by walls and disclinations (the so-called oily streaks). These textures are shown in Figure 3.4:

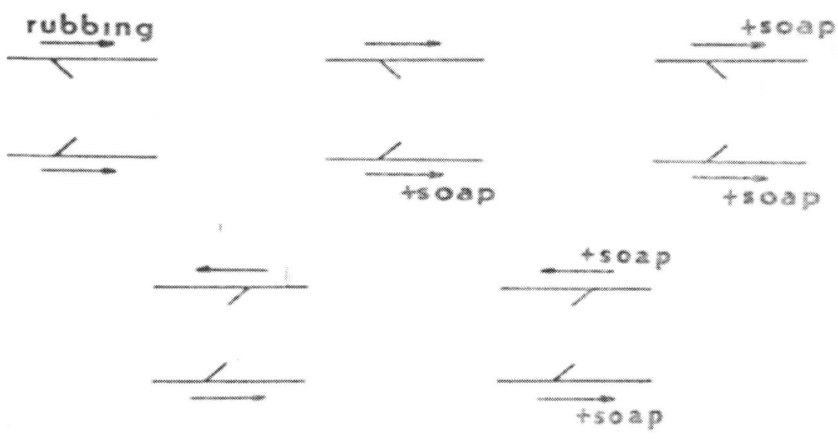

Figure 3.3. A schematic representation of the surface orientation of a large-pitch cholesteric-nematic mixture for all cases considered in the text.

The stripes show a small tilt of the cholesteric planes. They strongly depend on the mutual position of the two nicols.

c. Reversely-tilted Weak-Weak Anchored MBBA-5CB-CC Cholesteric Layer

Grandjean-like textures were observed at such an initial orientation, which were with a small tilt of the cholesteric planes of the large-pitch cholesteric expressed by the appearance of band domains shown in Figure 3.4.

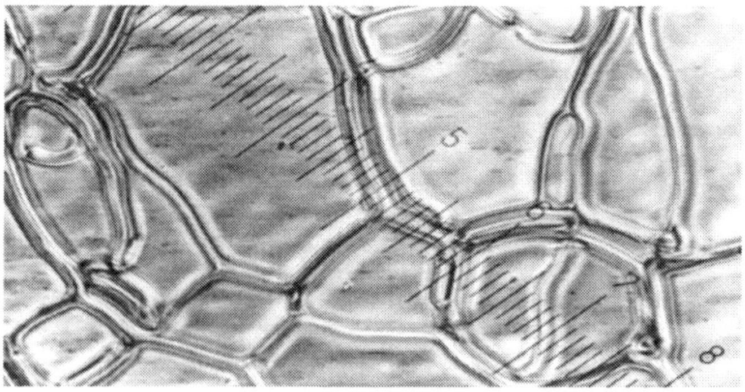

Figure 3.4. Large regions of Grandjean divided by "oily streaks" in Reversely-tilted strong-weak anchored MBBA-5CB-CC large-pitch cholesteric layers with a tickness of 15 μm,10 small divisions correspond to 45 μm; P⊥A, room temperature. The stripes into the regions show the weak tilt of the cholesteric layers and depend strongly on the mutual position of the two nicols.

As noted above, the view of the bands depends strongly on the mutual position of the two nicols. For example, if the polarizer is rotated with 40^0, one observes the image shown in Figure 3.5.

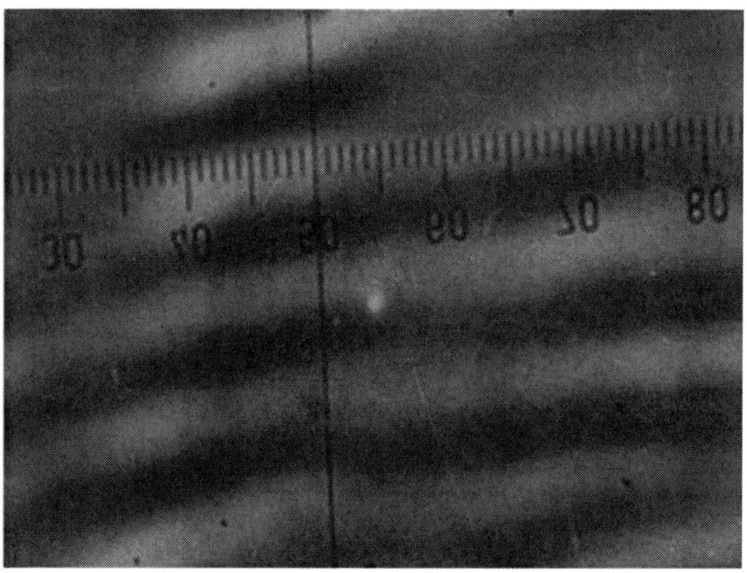

Figure 3.5. A Grandjean structure of the large-pitch MBBA-5CB-CC cholesteric with a thickness of 15 μm at Reversely-tilted weak-weak anchored large-pitch cholesteric layers.10 small divisions correspond to 45 μm; P50°A. Room temperature.

3.3. Electro-Optic Behavior of Strong-Strong, Strong-Weak and Weak-Weak Anchored Tilted Large-Pitch Cholesteric Layers of MBBA-5CB-CC Mixture

The electric field-induced textures and the phase-change behavior of the large-pitch cholesteric films are summarized in Table1.

Table 1. Texture and phase of LC films as a function of applied voltages and surface anchoring

Boundary conditions	$V = 0$	$\uparrow V > 3U_{th}$ $\uparrow V = V_t^{24}$	$V_{quenched}$	$\uparrow V > U_{sat}$	$V_{quenched}$
Reversely-tilted strong-strong anchored Ch layers	FPT_N		FPT_N	NT	FPT_N
Reversely-tilted strong-weak anchored Ch layers	TG + disclin.		FCT	NT	FCT
Reversely-tilted weak-weak anchored Ch layers	TG	Helfrich's instability	TG	NT	FPT TG($V < 3U_{th}$)
Tilted strong-strong anchored Ch layers	Nonoriented FPT		Nonoriented FPT_N	NT	Nonoriented FPT_N
Tilted weak-weak anchored Ch layers	Oriented FPT		Oriented FPT_N	NT	Oriented FPT_N
Homeotropic: Lin-Hendel[24]	ST/G	FCT	FCT	NT	ST
Homogeneous: Lin-Hendel[24]	G	FCT	FCT	NT	FCT

FPT — fingerprint texture; G — Grandjean's plane texture; NT — no texture; TG — tilted Grandjean's plane texture; FPT_N — fingerprint texture embedded in a homeotropic matrix; FCT — focal-conic texture; ST — scroll texture

For convenience, the corresponding electric field-induced textures, obtained by Lin-Hendel in planar and homeotropic cholesterics [51, 52] are given as well. The table shows that in our experiment the following texturtes have been observed: "fingers-print", Grandjean-like orientation, tilted Grandjean-like orientation, without a texture, "fingers print" embedded into a homeotropic matrix, focal-conic texture and a scroll texture. Our experimental results clearly show that the value of the voltage threshold leading to the unwinding of the "fingers" embedded into a pseudo-homeotropic nematic is the smallest one: in the range of 1 $V_{r.m.s.}$ for the case of reversely-tilted surface orientation and usual treatment (only rubbing) of the glass plates (let us remind that $\Delta\varepsilon$ of our mixture is between 1.5 and 2; if $\Delta\varepsilon$ is in the range of 15-20, the voltage threshold should be in the millivolt region). The further increase of the voltage led to further unwinding of the large-pitch cholesteric. It is important to note that the "fingers" with a pitch of 12 μm disappeared in a homogeneous as shown in Figure 3.6 and Figure 3.7.

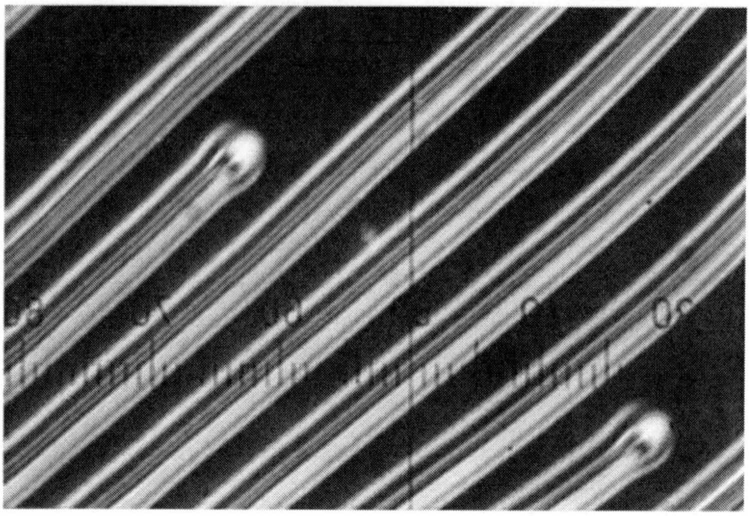

Figure 3.6. Fingerprint texture embedded into a homeotropic matrix in large-pitch MBBA-5CB-CC tilted cholesteric with a strong-strong anchoring and a thickness of 15 μm. 10 small divisions correspond to 45 μm. An AC voltage of $2V_{rms}$ and a frequency of 10 kHz was applied to the layer; room temperature, P⊥A.

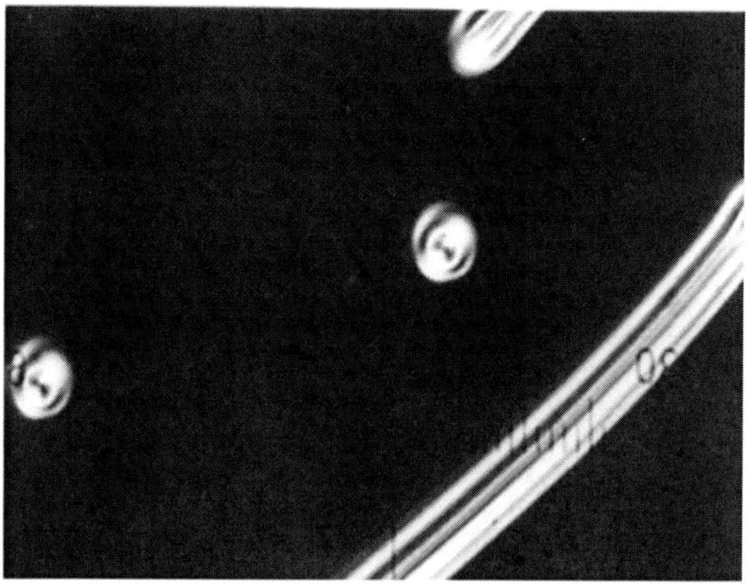

Figure 3.7. View at an AC voltage of $3V_{rms}$.

(U = 3 $V_{r.m.s.}$, respectively U = $4V_{r.m.s.}$). At decrease of the voltage, however, one observes non-homogeneous and non-linear appearance of the "fingers" as shown In Figure 3.8 and Figure 3.9.

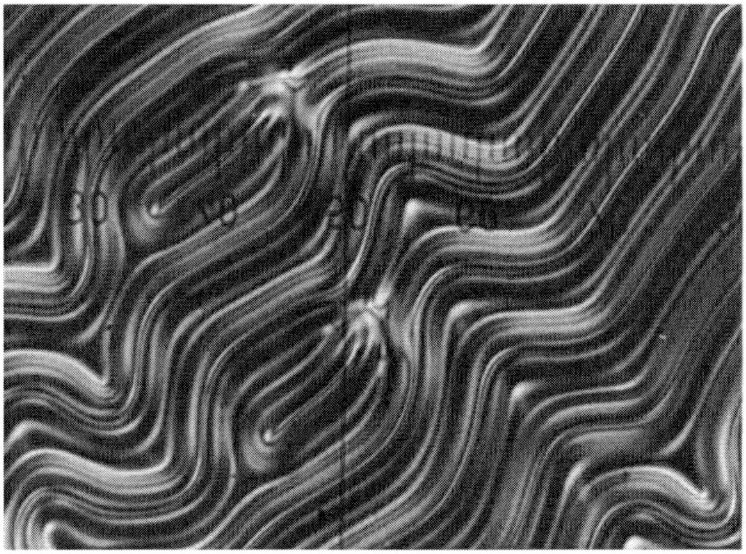

Figure 3.8. "Hering-bone" texture of Rault-Cladis at descreasing of the AC voltage (f=10 kHz) from 3 $V_{r.m.s.}$ to 2 $V_{r.m.s.}$.

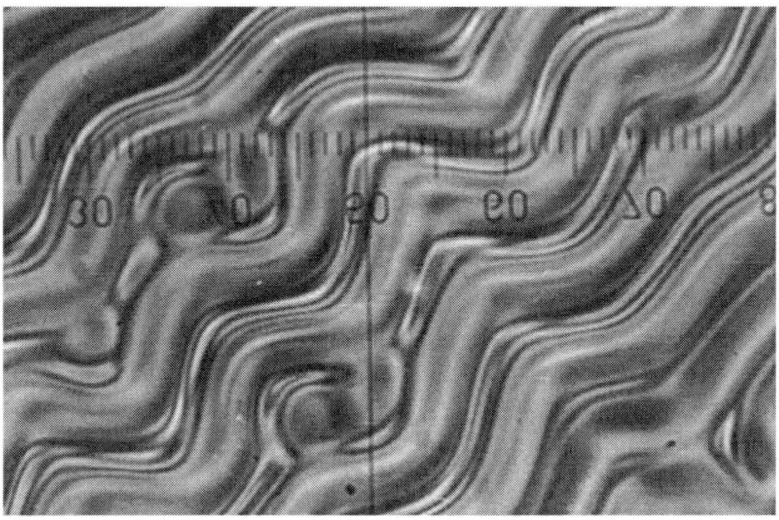

Figure 3.9. "Hering-bone" texture of Rault-Cladis at descreasing of the AC voltage from $2V_{rms}$ to $1V_{\mathbf{rms.}}$ (f=10 kHz).

for a voltage of $2V_{r.m.s.}$ and 1 $V_{r.m.s.}$, respectively. Electro-optical transmission curves for the case of a finger-print texture, embedded into a homeotropic matrix (U = 4 $V_{r.m.s.}$, U_c=$2V_{r.m.s.}$, curve 3) is shown in Figure 3.10 for a reversely-tilted strong-weak anchored MBBA-5CB-CC mixture 15 μm thick at room temperature. After an abrupt disconnection of the voltage, the liquid crystal returns to its equilibrium state, i.e., such textures do not show any hysteresis and bistability. As a further step, we will show the electro-optic behavior of a liquid crystal cell with confocal texture at a switch-on (Figure 3,11, U_c= 1 $V_{r.m.s.}$, d=15 μm) or at switch-off of the voltage on (Figure 3,12,) and after 10 minutes (Figure 3.13) or at switch-off of the voltage on (Figure 3,12,) and after 10 minutes (Figure 3.13).

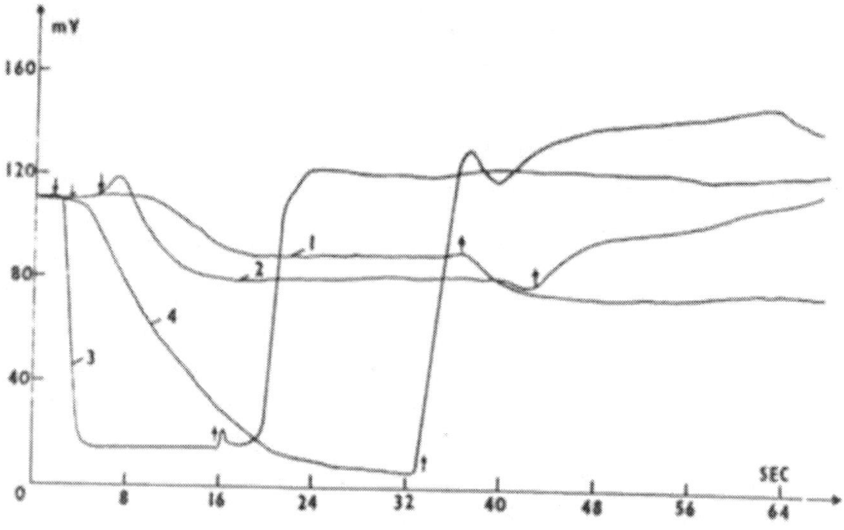

Figure 3.10. Electrooptical transmission curves for reverse-tilted strong-weak(curve 1, $U=10V_{r.m.s.}$,$U_c=5V$), weak- weak (curve 2, $U=4V_{r.m.s.}$, $U_c=2V$), weak- weak(curve 4, $U=15V_{r.m.s.}$, $U_c=2V$) anchored MBBA- 5 CB- CC mixture 15 μm thick at room temperature. The curve 3 corresponds to the case of a „finger- print"texture embedded into a pseudo- homeotropic matrix of the same conditions: $U=4\ V_{r.m.s.}$, $U_c=2$ V. P‖A.

Figure 3.11. Focal- conic texture in planar MBBA- 5 CB- CC mixture, 15 μm thick, at application of AC voltage of 1 V_{rms}(f=10 kHz). 10 small divisions correspond to 45 μm. P⊥A.

Textures, similar to those shown in Figure 3.12 and Figure 3.13 have been observed by Kashnow et al., at relaxation of pseudo-nematic towards the cholesteric texture [70]. The electro-optic behavior of Grandjean-like orientation with strong-weak anchoring, expressed by a number of textures [71], is shown for example in Figures 3.14 to Figure 3.19.

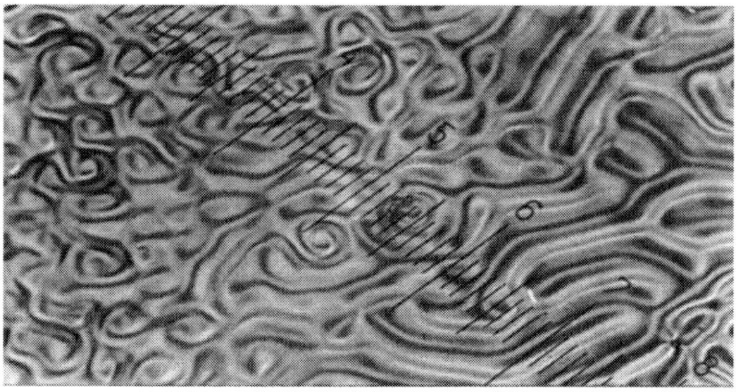

Figure 3.12. A view after the switch-off of the applied voltage.

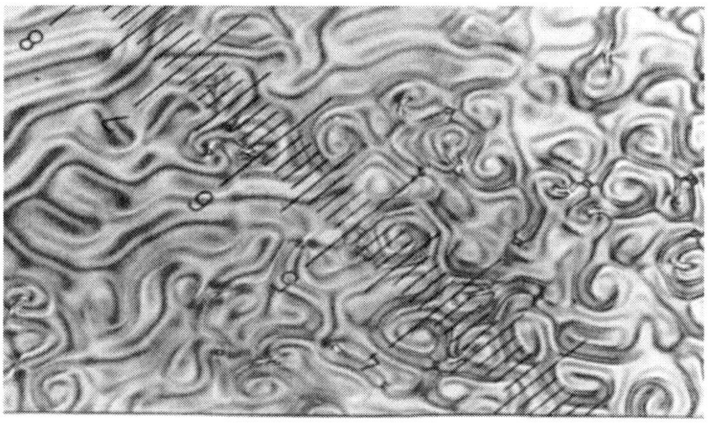

Figure 3.13. A view after 10 minutes of the switch-off of the applied voltage.

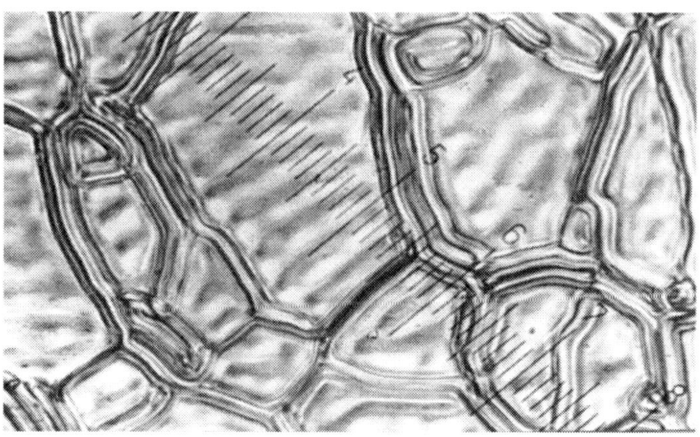

Figure 3.14. The texture of Figure 3.5 at application of AC voltage with an amplitude of 0,5V_{rms} (f= 10 kHz). {It is seen the appearance of Helfrich- Hurault instability. 10 small divisions correspond to 45 μm. Room temperature, P⊥A.

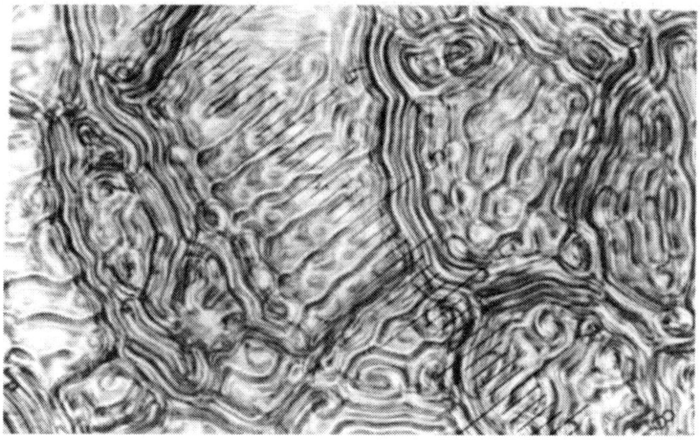

Figure 3.15. A view at applications of AC voltage with an amplitude with $1V_{rms}$ (f=10 kHz).

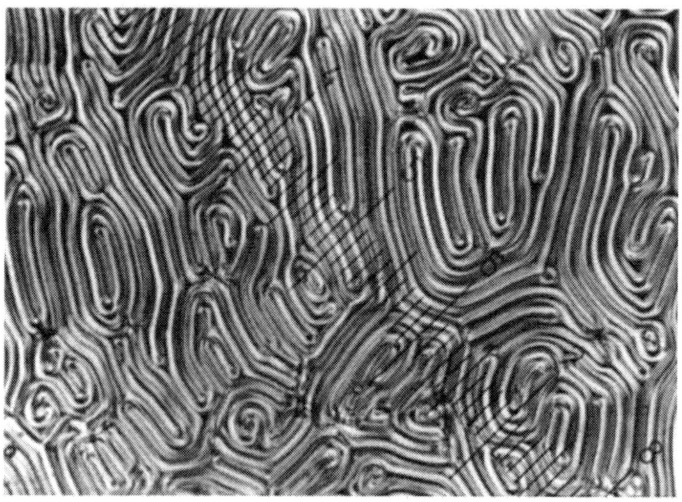

Figure 3.16. A view after application of AC voltage with an amplitude of $3V_{rms}$ (f=10 kHz).

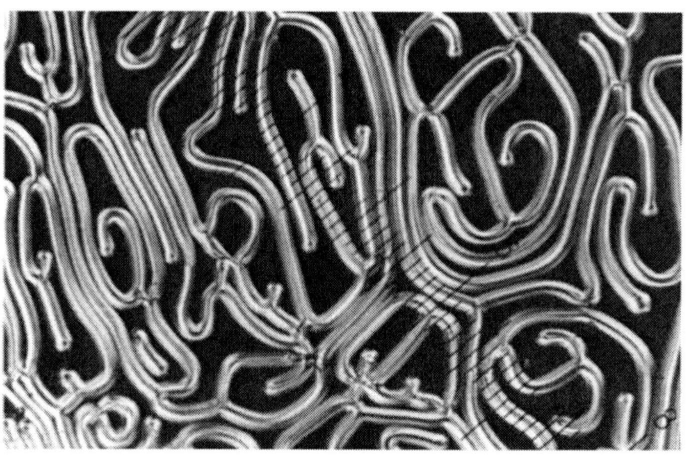

Figure 3.17. Beginning of un-twisting of the focal-conic texture at application of AC voltage with an amplitude of 4 V_{rms}(f=10 kHz).

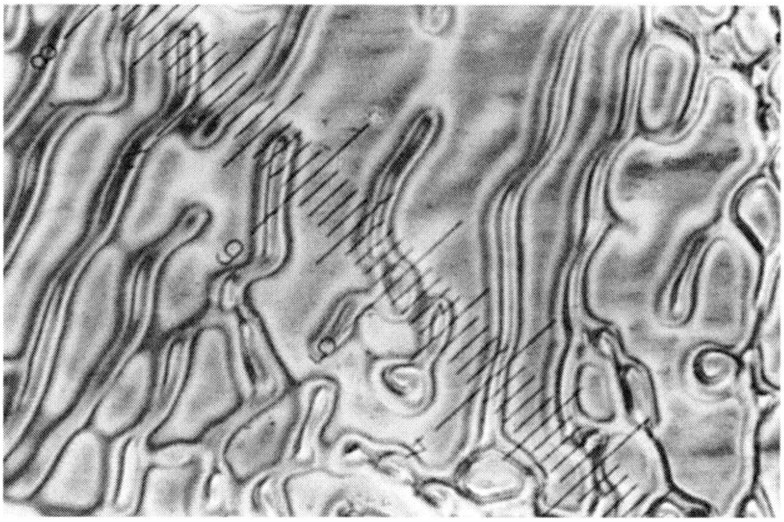

Figure 3.18. A view after 5 minutes of the switch-off of the applied voltage.

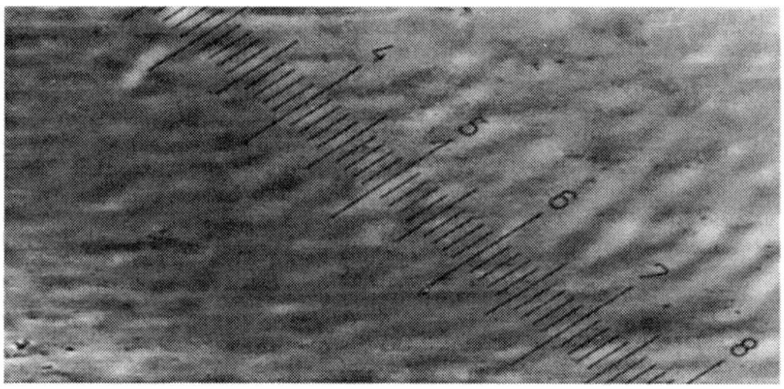

Figure 3.19. The change in the picture after 10 minutes.

The instability of Helfrich-Hurault appears first at a voltage of 0.5 $V_{r.m.s.}$ (Figure 3.14) followed by the formation of focal-conic and finger-print textures at raising of the voltage to 1 $V_{r.m.s.}$(Figure 3.15). Further rise of the voltage to 2 $V_{r.m.s.}$ led to the formation of the texture, shown in Figure 3.16 which rest unchanged although the voltage was raised from 2 to 3 $V_{r.m.s.}$. The beginning of the unwinding of the large-pitch cholesteric textures occurs at a voltage of 4 $V_{r.m.s.}$(Figure 3.17). The relaxation of the pseudo-homeotropic nematic towards the cholesteric texture after 5 minutes and after 10 minutes is shown in Figure 3.18 and Figure 3.19, respectively. The electro-optic transmission curve is shown for a reversely-tilted strong-weak, as noted, in Fgure 3.10 (curve 1, U = 10 $V_{r.m.s.}$, U_c = 5 $V_{r.m.s.}$, f = 1 kHz). The electro-optic transmission curves for weak-weak (curve 2, U = 4$V_{r.m.s/}$, U_c=2$V_{r.m.s.}$) anchored MBBA-5CB-CC mixtures are also shown in Figure 3.10. The Helfrich-Hurault instability is shown in Figure 3.20 to Figure 3.22.

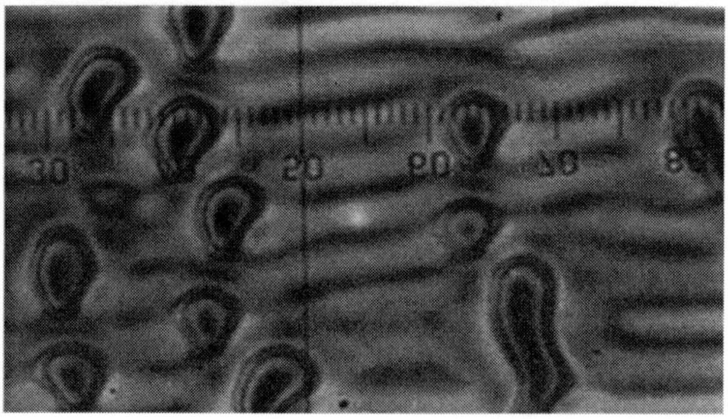

Figure 3.20. The instability of Helfrich- Hurault in weak- weak anchored MBBA-5CB-CC large-pitch cholesteric with initially Grandjean-like texture under AC voltage of 2 V_{rms} (f=1 kHz), d=15 μm, room temperature. 10 small divisions correspond to 45 μm, P⊥A.

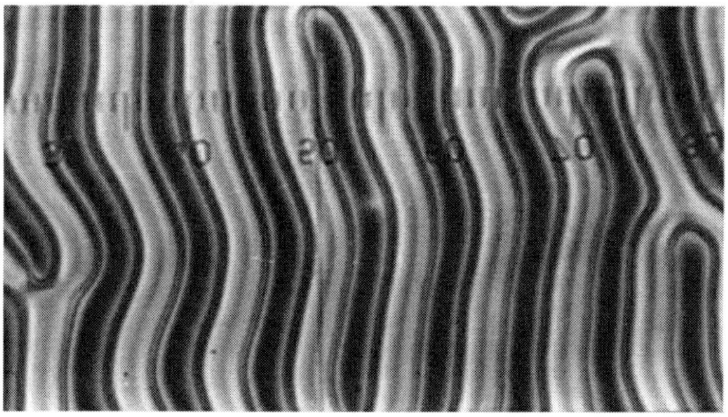

Figure 3.21. A view at application of AC voltage with an amplitude of 2,1 V_{rms}(f=1 kHz).

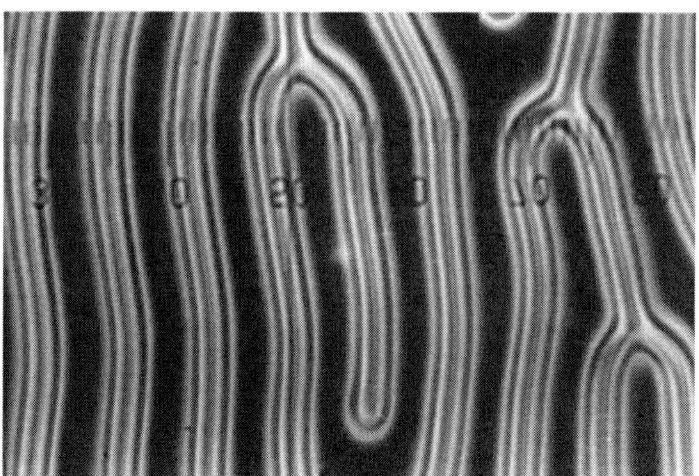

Figure 3.22. π/2 rotation of the cholesteric axis and appearance of finger-print embedded into a pseudohomeotropic nematicat \ volt\ge of 2.7 $V_{r.m.s.}$.

Figure 3.20 shows the appearance of Helfrich-Hurault instability at 2.0 $V_{r.m.s.}$, f = 1 kHz. Figure 3.21 shows the development of this instability at a voltage of 2.1 $V_{r.m.s.}$ and Figure 3.22 shows the appearance of the texture finger-print embedded into a pseudo-homeotropic nematic at a voltage 2.7 $V_{r.m.s.}$(let us stress that the large-pich cholesteric is with a reversely-tilted weak-weak-anchored boundary layers). It is remarkable that the decrease of the voltage from 6 $V_{r.m.s.}$ to 2.1 $V_{r.m.s.}$ leads to the reappearance of the Helfrich-Hurault domains with additional undulation of some of the cholesteric layers in a perpendicular direction. This picture is shown in Figure 3.23.

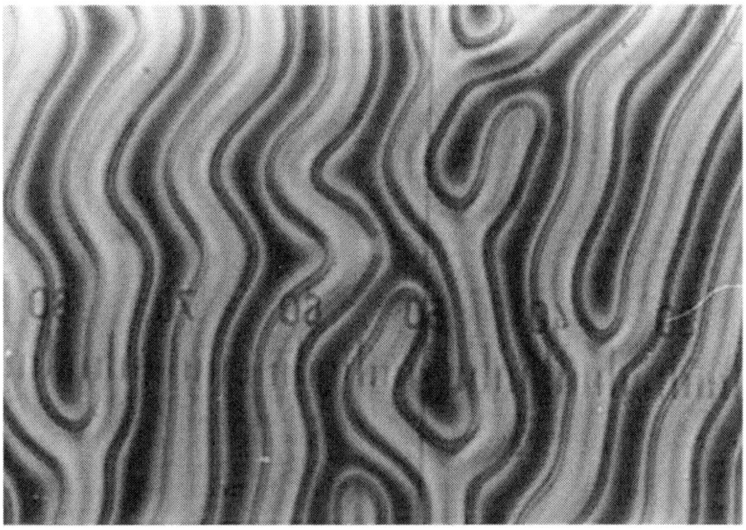

Figure 3.23. A new $\pi/2$ rotation of the cholesteric axis and re-appearance of Helfrich-Hurault instability with undulations of the cholesteric layers in some places at decrease of the voltage from 6 $V_{r.m.s.}$ to 2.1 $V_{r.m.s.}$.

3.4. Hysteretic Electroptic Behavior of Large-Pitch Cholesterics. Hysteretic Behavior of Reversely-Tilted Weakly-Anchored Large- Pitch Cholesterics

Many authors studied the dynamic electro-optic behavior of large-pitch cholesterics showing hysteresis and bistability. In our experiment [72], the large-pitch cholesterics were excited by a linear (more precisely, triangular) changing the amplitude of the voltage with a frequency of 10 kHz. In effect, Gerber [53], and Schadt and Gerber [34] have used this method. We know that the liquid crystal cannot immediately follow the fast changes of the voltage due to the bulk and surface rotational viscosities. The hysteresis depends strongly on the kind of the scanning voltage (linear, rectangular, etc) and on the scanning velocity of the voltage. The scanning with a velocity of 0.02 V/sec of a large-pitch cholesteric with initial Grandjean orientation and weak surface anchoring led to a decrease of the voltage, to the appearance of an instability of Helfrich-Hurault or to the domains of Rault-Cladis [73] and finally to the Fréedericksz transition which shows the unwinding of the cholesteric. Inversely, at increase of the voltage, scanning the cholesteric with a 0.2 V/sec velocity, the domains of

Rault-Cladis disappear first and the domains of Helfrich-Hurault disappear at a higher scanning velocity of 2V/sec. Similar hysteretic curves were obtained for the texture "finger-print" embedded into one pseudo-homeotropic nematic when we did not use the soap. The comparison of all hysteretic curves for this case shown in Figure 3.24 clearly pointed out that the bigger hysteresis, and consequently bistability, are obtained with a "finger-print" texture, embedded in a pseudo-homeotropic nematic and scanned with a velocity larger than 2V/sec.

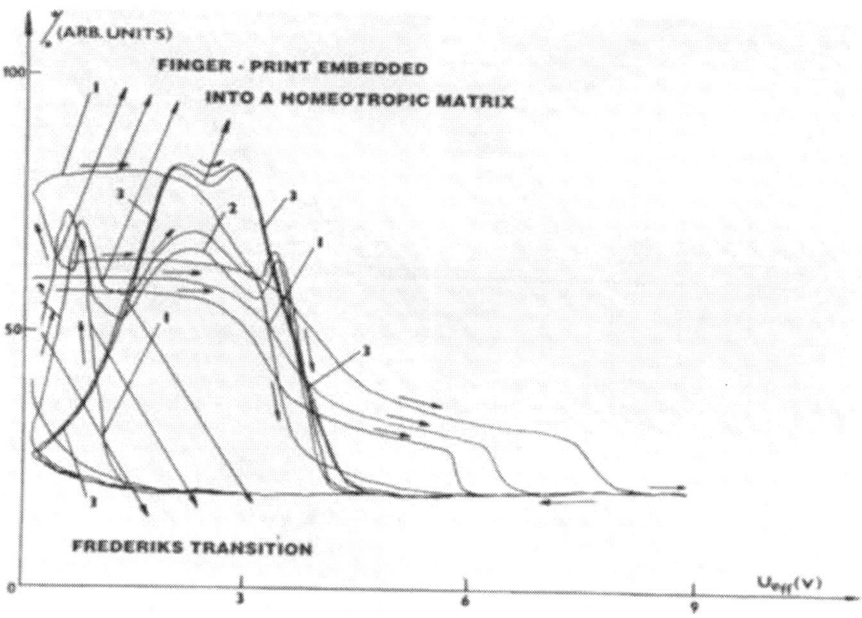

Figure 3.24. Dynamic hysteretic curves recorded after the application of AC voltage with a frequency of 10 kHz, linearly scanning the large-pitch cholesteric MBBA-5CB-CC mixture with a pitch of 12 μm and a thickness of 15 μm. The rubbing direction is along the bisectris of the crossed nicols. The initial texture is"finger-print" embedded into a pseudo-homeotropic matrix. The curves designated by 1, 2 and 3 correspond to a sweeping rate of the voltage of 1V/sec, 2V/sec and 3V/sec. The arrows indicate the sweeping direction of the voltage.

The unwinding of the spirals is effected at a much weaker voltage (in comparison with that used by Gerber and by Schadt and Gerber) due to the weak surface anchoring of the large-pitch cholesteric, especially for the Grandjean orientation. It is undisputable that the weak surface coupling increases the hysteresis. In this case, however, the participation of the surface rotational viscosity, the inverse flow in the bulk and in the surface regions and the surface tilt are of importance as well. The scanning of the voltage with a slower velocity permits the observation of the Rault-Cladis domains and the conical deformation of the cholesteric, which is usually not observable due to the influence of the various textures. The dynamic hysteresis can find application in the liquid crystal matrices with storage, and more exactly in their electrical control [74-78]. In summary, the boundary orientation of cholesteric director and the surface energy [79-82] are of great importance for the observation of various kinds of cholesteric textures and their behavior under applied AC electric fields.

3.5. Unusual Electro-Optic Behavior of a New Chiral Mixture of Nematic and Ferroelectric Smectic-C*

The large-pitch cholesterics used in the scientific experiments and liquid crystal optoelectronics are prepared usually by mixing a nematic liquid crystal with a small amount of several weight percent of a short-pitch cholesteric or a nonmesomorphic chiral additive. Uemoto et al., [83] have investigated a mixture of the ferroelectric smectic C* liquid crystal DOBAMBC, prepared for the first time by Meyer et al., [84] with the liquid crystal PAA in order to obtain a change in the pitch of twisting of the ferroelectric liquid crystal, its spontaneous polarization, the dielectric properties of the mixture and the temperature interval of existence of the ferroelectric smectic C* liquid crystal. According to these authors, the small organic molecules of the liquid crystal PAA are located among the molecules of DOBAMBC and decrease the intermolecular interactions via the increase of the freedom of rotation around their long axes. From a chemical point of view, they supposed that the molecules of PAA are located around the Schiff bases of the molecules of DOBAMBC. Uemoto et al., found that 25 wt. % of PAA is sufficient to remove the ferroelectric properties of the PAA-DOBAMBC mixture. In this part of the chapter, we will describe the textural, optical and electro-optical behavior of a mixture of the ferroelectric liquid crystal DOBAMBC with the nematic phase of the liquid crystal PCH-7. The mixtures of 1-5 wt. % DOBAMBC/99-95 wt. % PCH-7 show usual textures typical for the conventional large-pitch cholesterics. However, the mixtures of 10-50 wt. % DOBAMBC/90-50 wt. % PCH-7 show different, in our opinion unusual behavior. The Grandjean structure, also in our opinion, shows unusual optic behavior and very intensive colors depending on the mutual position of the two nicols. The very fast reaction of the liquid crystal cell at the unwinding of the cholesteric which is in the range of several milliseconds also show the "unusual" electro-optic behavior of these mixtures. Some of these mixtures are very promising for application in the liquid crystal displays working with large-pitch cholesterics. The material parameters of the ferroelectric smectic-C* liquid crystal are well- known This liquid crystal has the following phases and phase transition temperatures:

$$DOBAMBC : I_{117°C} - S_m A_{95°C} - S_m C^*_{76°C} - C \tag{3.14}$$

at heating and

$$DOBAMBC : I_{117°C} - S_m A_{95°C} - S_m C^*_{63°C} - S_m H_{45°C} - C \tag{3.15}$$

at cooling and the following chemical formula[84]:

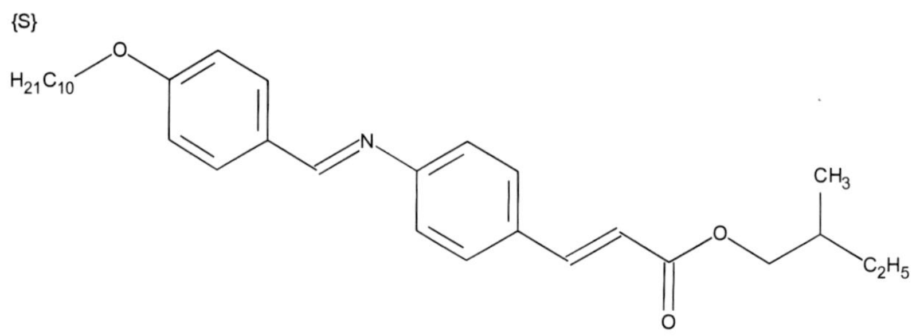

The seventh homologue of the well-known liquid crystal series of p-cyano-p'-alkyl phenyl cyclohehexane (PCH-7) [85-91] has the following phase transitional temperatures

$$PCH-7: I_{60^0 C} - N_{30^0 C} - C \tag{3.16}$$

and the following chemical formula:

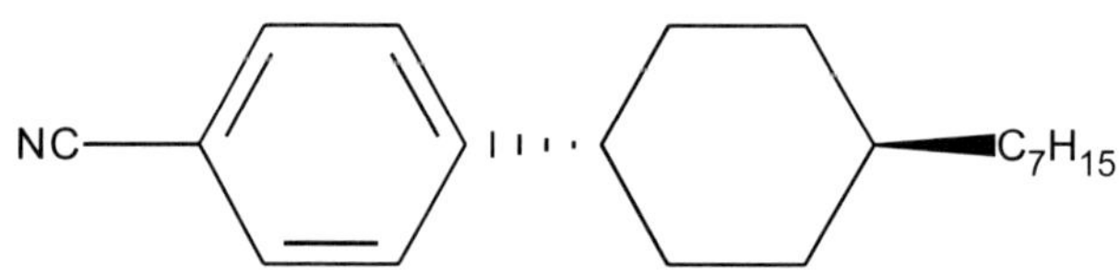

Additionally we have studied the following mixtures: DOBAMBC/8CB, DOBAMBC/PCH-5 and DOBAMBC/ LC 1132 (this liquid crystal has a positive dielectric anisotropy of 19.3 at room temperature).

The material parameters of the ferroelectric smectic-C* liquid crystal DOBAMBC are well known [92-98]. The two dielectric anisotropies are negative and their magnitude is around 1. The other material parameters can also be found in the literature. It is interesting to note that the rotational viscosity of the ferroelectric smectic-C* DOBAMBC is of the same order of magnitude as that of the nematics, the elastic constants of splay and bend are bigger than those of the nematics whereas the elastic constant of twist is of the order or even smaller than that of the nematics. Since the magnitude of the rotational viscosity of the ferroelectric smectic-C* liquid crystal is of the same order as that of the nematic, the range of K_{22} can be found from the time reaction of the electro-optic effects of different mixtures.

We have prepared seven DOBAMBC/PCH-7 mixtures. The glass plates of the liquid crystal cells were preliminary rubbed with abrasive paste in order to achieve planar orientation of the liquid crystal director. Two polymer sheets with a thickness of 8, 10 and 20 μm have determined the thickness of the liquid crystal cells. Initially, a small drop of the liquid crystal PCH-7 was added laterally to the isotropic phase of DOBAMBC placed in the cell. This well- known "gradient" method permits to observe easily all possible liquid crystalline phases which are formed under the concentration gradient of PCH-7. The liquid crystal cells were placed in the Mettler FP 82 hot stage. Mettler FP 800 central processor ensured the electronic regulation of the temperature. The liquid crystal mixtures were observed under polarizing microscope in transmitted white light. Four liquid crystals were formed: nematic, new chiral phase, smectic-A and ferroelectric smectic-C* phase. We have also studied the DOBAMBC/PCH-5 and DOBAMBC/LC 1132 mixtures and observed the same liquid crystals. The new chiral phase, however, was not observed in the DOBAMBC/8CB mixture. The replacement of the more active PCH-R liquid crystal with the less-active 8CB led to observation of only two liquid crystals: smectic-A and smectic-C* phases. Further, we have prepared seven DOBAMBC/PCH-7 mixtures (Figure 3.25) as follows:

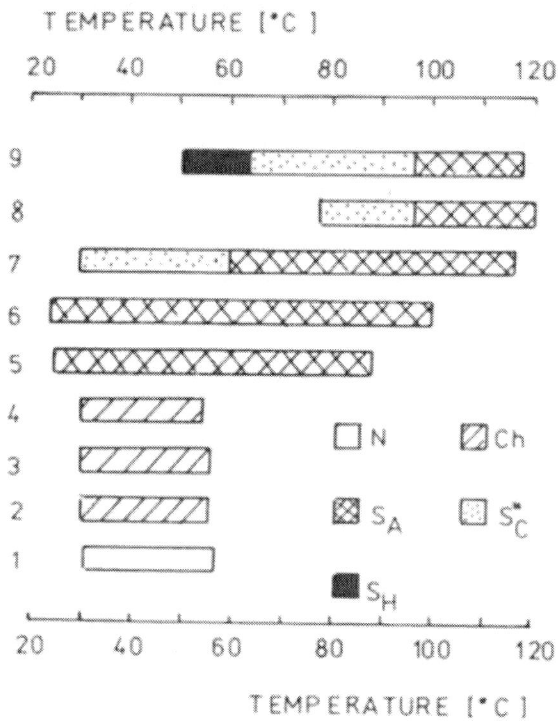

Figure 3.25. A temperature range of the liquid crystal phases of PCH-7 (1), mixture A to D(2), mixture E(3), mixture F(4), mixture G(5), mixture H(6), mixture E(7). DOBAMBC at heating (8) and DOBAMBC at cooling (9).

(with (1) is designated the liquid crystal PCH-7, with (2) – the mixtures from A to D (5-25 wt. % DOBAMBC), with (3) – the mixture E (35 wt. % DOBAMBC), with (4) – the mixture F (50 wt. % DOBAMBC), with (5) – the mixture G (65 wt. % DOBAMBC), with (6) – the mixture H (80 wt. % DOBAMBC), with (7) – the mixture I ((90 wt. % DOBAMBC), with (8) – DOBAMBC at heating and with (9) – DOBAMBC at cooling (the rest wt. % to 100 are of the liquid crystal PCH-7)). All these mixtures showed different textures from conventional cholesteric textures (a mixture of 5 % wt.% DOBAMBC – 95 wt. % PCH-7), which were replaced by smectic-like textures such as paramorphic mosaic texture and fan-texture. The increase of the DOBAMBC between 65 and 80 wt. % led to the disappearance of the cholesteric and to the appearance of the smectic-A phase. The increase of content of DOBAMBC (for instance 90 wt. %) led to the appearance of the ferroelectric smectic-C* phase at considerably low temperature. Our experiment demonstrated that the cholesteric phase does not show ferroelectricity in the frame of the experimental error of 0.1 nC/cm^2.

Some of the mixtures showed remarkable color properties. For example, the cell with a mixture of 20 wt. % DOBAMBC – 80 wt. % PCH-7, 20 μm thickness and Grandjean orientation showed very intensive dark green, wine-red and dark yellow colors at different mutual position of the two nicols.

3.6. ELECTRO-OPTIC BEHAVIOR OF THE LIQUID CRYSTAL MIXTURES

Under application of AC voltage, the electro-optic behavior of the mixtures studied by us was complex with initial cholesteric textures [99] – [103], their transformation at raising of the voltage in other textures, non-typical for the cholesterics and finally, at higher voltage, the cholesteric was totally unwinded. We would like to illustrate this with the study of Asai et al., [104] who have observed the unwinding of the long-pitch cholesteric under the action of AC electric field. They have observed the following textures: confocal texture-intermediate confocal texture- polygonal texture –finger print texture-pseudo-homeotropic nematic when the initial confocal texture was exposed to AC voltage and the cholesteric is with a positive dielectric anisotropy. It is evident that the cholesteric layers in the confocal and polygonal textures are considerably bent which depends on the value of $(2\pi/p_0)^2K_{22}$ and $(3/8)K_{33}$ which describe the change in the amplitude of bending of the cholesteric layers along Z and X, respectively, according to the relation, given by Scarramuza et al., [105]:

$$F(x,z)=\left(\frac{1}{2}\right)\left(\frac{2\pi}{p_0}\right)^2 K_{22}\left(\frac{\partial u(x,z)}{\partial z}\right)^2+\left(\frac{1}{2}\right)\left(\frac{3}{8}\right)K_{33}\left(\frac{\partial^2 u(x,z)}{\partial x^2}\right)^2 \tag{3.17}$$

where F(x, z) is the local density of the elastic energy of the cholesteric, given by de Gennes [40]. It is clear that the cholesteric layer will be hardly bent for the case of small pitch and large elastic constant of twist and bend and inversely, the cholesteric layer will be easily deformed for large pitch and small elastic constants of twist and bend.

The electro-optic behavior of the mixtures, considered here, was complex. Formally it could be divided into three parts: electro-optic behavior of mixtures containing more than 1 wt. % DOBAMBC, electro-optic behavior of mixtures containing between 10 wt. % and 25 wt. % DOBAMBC and finally, electro-optic behavior of mixtures containing between 25 wt. 5 and 50 wt. % DOBAMBC. At the first class textures, the Grandjean-like texture and paramorphic texture at heating and confocal texture at cooling were the initial stable textures. They were transformed into polygonal texture at application of a voltage with an above-threshold value (see Figure 3.26).

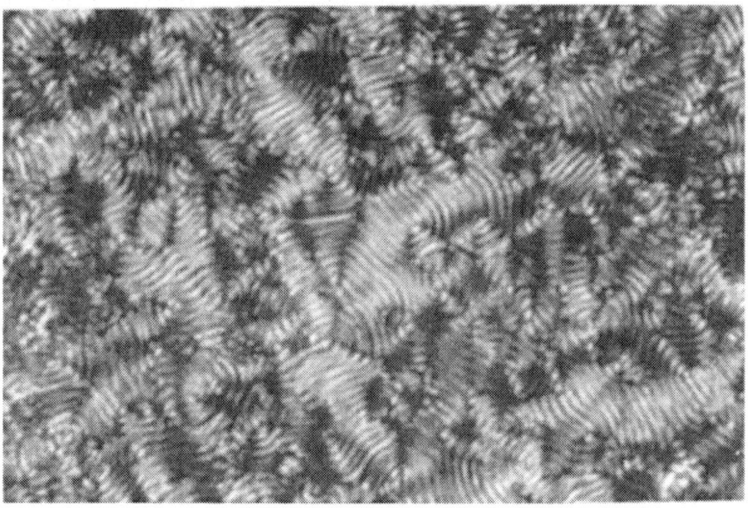

Figure 3.26. Electro-optic textures in 10 wt %_DOBAMBC- 90 wt % PCH-7(mixture C) under a voltage above 3 V_{rms}(U_c=3 V_{rms}, d=8 μm, P⊥A, room temperature, the long side of the picture corresponds to 485 μm). They represent polygonal textures with twisting of the cholesteric large-pitch layers in the planes of the Electrodes.

(Let us note that there is a difference between the voltage, showing the beginning of the unwinding of the spiral, and the voltage showing the complete unwinding: 3 $V_{r.m.s.}$versus 11 $V_{r.m.s.}$). At a sudden removal of the voltage, one obtains a transitional Grandjean-like texture with a pitch of $p_0(K_{33}/K_{22})$ which slow relaxes toward the usual Grandjean-like texture and in some places with the formation of "scroll" texture due to the bad adaption between the orientation of the molecules at the boundaries and the Grandjean-like texture. At the second class of mixtures, the initial texture is fan-shaped with stripes (see Figure 3.27) by the application of a voltage with above-threshold amplitude. At higher voltage, one observes complete unwinding of the cholesteric and formation of pseudo-homeotropic nematic.

Figure 3.27. The texture from Figure 3.26 at an application of voltage of 4 V_{rms}.

At the third class of mixtures, when DOBAMBC is from 25 wt. % to 50 wt. %, the initial texture is Grandjean-like which under the application of a voltage with above-threshold value goes to fan-shaped and finally, to complex polymorphic-chevron texture (see Figure 3.28).

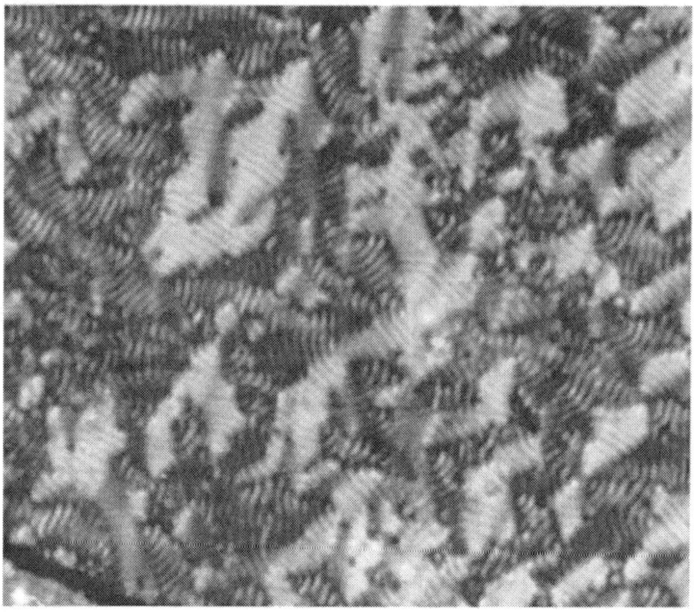

Figure 3.28. Electro-optic textures in 35 wt %_DOBAMBC- 65 wt % PCH-7 (mixture F) under a voltage above $6V_{rms}$, d=8 μm, P⊥A, room temperature, the long side of the picture corresponds to 485 μm. The pitch of twisting decreases with increase of DOBAMBC.

The stripes in the textures, showing the pitch of the large-pitch cholesteric decreased from 6 μm to 2 μm with rthe increase of the amount of DOBAMBC. The behavior of the cholesteric planar texture under the action of external electric field has been considered by many authors [106-108].

We will focus here our attention on the electro-optic behavior of a cholesteric material, which can be used in the electro-optic devices for modulation of the light, and more specifically for a reproduction of digits, signs and images [109]. This liquid crystal mixture shows a high contrast and a high velocity of relaxation at application of a voltage with above-threshold value (a high electric field) and its removal. The relaxational times at the cholesteric-nematic and at the nematic-cholesteric phase transition are from 5 to 10 times smaller relative to those of the cells with conventional large-pitch cholesteric. Such short relaxation times are peculiar for the mixtures containing the ferroelectric smectic-C* DOBAMBC from 5 to 25 wt %. In order to achieve such relaxation times, the pitch of the mixture must be equal to or greater than 1 μm, the dielectric anisotropy must be positive and the initial texture of the mixture (in the moment of the application of the voltage) can be either Grandjean-like or confocal, the nicols should be crossed. At these conditions, one obtains electro-optical transmission curves shown in Figure 3.29 (the applied voltage should be with amplitude greater than 10 V (curves 2 and 3 in Figure 3.29. The time reactions are very small in the range of several milliseconds (Figure 3.30 and Figure 3.31).

These times are commensurate with the best ones measured in the displays developed on the basis of the cholesteric-nematic phase transition[110-116] and slightly higher than those measured with nematics [117, 118].

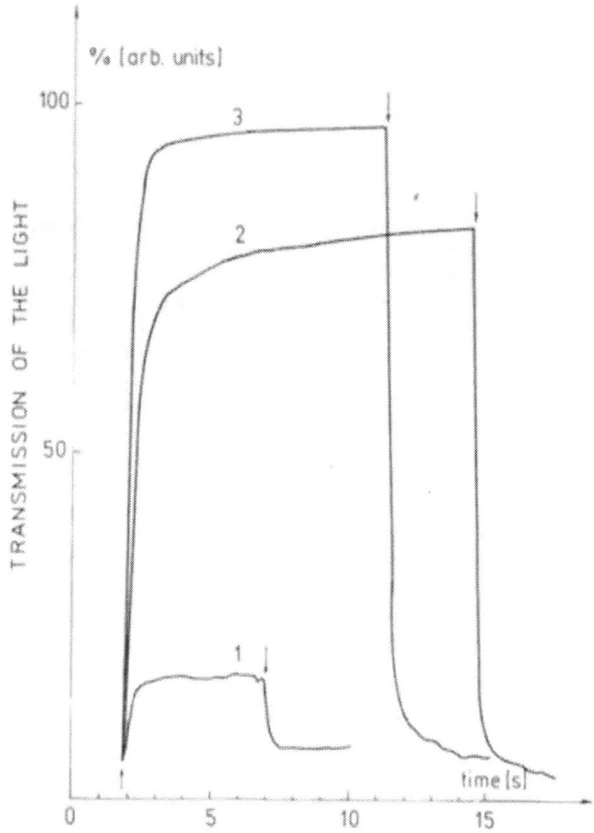

Figure 3.29. Electro-optic transmission curves obtained with ana electronic recorder for a 10 wt % DOBAMBC- 90 wt % PCH-7 mixture with a thickness of 8 μm at switch-on and switch-off of a voltage of $3V_{rms}$, 5 kHz (curve 1), 10 V_{rms} (curve 2) and $15V_{rms}$ (curve 3), P⊥A, T=29°C.

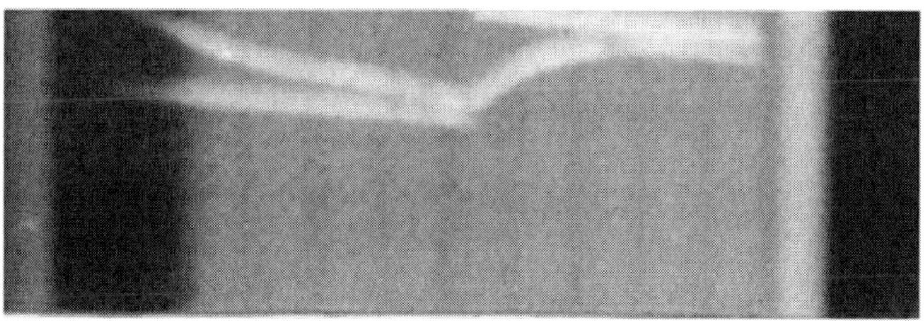

Figure 3.30. The reaction times from curves in Figure 3.29, shown on the monitor of the oscilloscope.

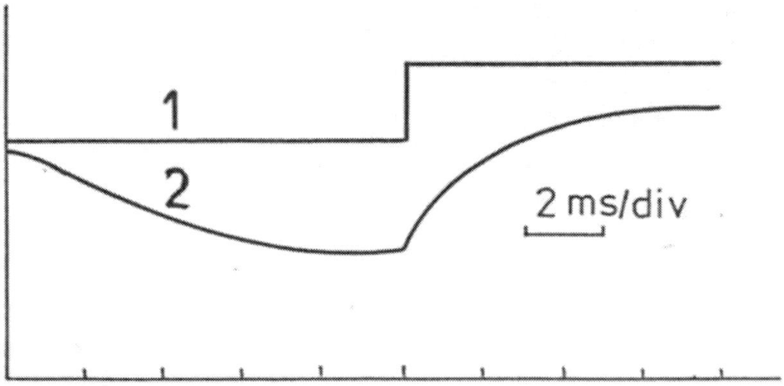

Figure 3.31. Redrawn curve from the oscillogram with transient rise and decay tmes for the mixture of 10 wt. % DOBAMBC – 90 wt. % PCH-7 with a thickness of 8 μm. The voltage impulse has a length of 10^5 μs, period of $2.5x10^4$ μs and a height of 10 V. The whole time scale is 29 ms with 2 ms per division.

Let us assume that the cholesteric is in initial confocal texture. At application of the voltage, the reaction times are determined by the following formulae:

$$\tau_{CN} = \eta / \varepsilon_a \left(E^2 - E_{CN}^2\right) \quad (3.18)$$

$$E_{CN} = \left(\pi^2 / p_0\right)\left(K_{22} / \varepsilon_a\right)^{1/2}$$
$$\tau_{NC} = \eta p_0^2 / K_{22} \pi^4 \quad (3.19)$$

It is seen, that the reaction times decrease when the value of the viscosity decreases (it is well-known that the nematic liquid crystal PCH-7 has a small viscosity), when the twist elastic constant K_{22} is large and when the value of the applied electric field is high. At the application of a voltage with a threshold determined by equation (3.4) the initial Grandjean texture is transformed into a confocal one and then raising the voltage (see (3.18)) into a pseudo-nematic.

3.7. Observation of Unidirectional Spirals in Nematic – Cholesteric Mixtures: A New Polar Electrooptic Effect in Thin Large-Pitch Cholesteric Films General Discussion

Two-dimensional spirals with both winding can be easily formed in the confocal textures, typical for the large-pitch cholesterics. Rondelez [119] has observed unidirectional spirals under the action of an electric field with a frequency of 6 Hz in such a way that one array of spirals with one winding have been situated near the one glass plate and another array of spirals with the opposite winding have been situated near the second glass plate of the liquid crystal cell and the two arrays of spirals pass each other with respect to the plane of the

Electrodes, respectively to the plane of the glass plates. Bouligand [99] pointed out that spirals can be formed only when the cholesteric layers are vertical or tilted. For this reason they can be easily formed in "finger-print" texture embedded or not embedded in a homeotropic pseudonematic as well as (as noted) in a confocal texture. Spirals can also exist in connection with various defects in long-pitch cholesterics [101] They can be observed in both dynamic or static systems. According to Kamien and Selinger [120], spirals can be easily formed in substances with a layered structure. They can be easily formed in large-pitch cholesteric drops which are posed on a glass plate and are surrounded by air. At these circumstances, the conditions are best for the formation of spirals. In this way [Saupe [65] has observed unidirectional spirals near the large-pitch cholesteric – air boundary. Unidirectional spirals were observed by us in 1983 [122]. On the basis of our experimental observations, two theories have been proposed: flexoelectric (or static) by Terentijev and Pikin [123] and dynamic, based on the theory of Ginzburg-Landau [124-128], or on the more general theory of Ginzburg-Landau-de Gennes [129–132] by Frish, Gil and Gilli [133]. We will see that all observed by us unidirectional spirals were created from fingers embedded in a homeotropic pseudo-nematic. The optical properties of these fingers have been considered in [134]. Ribiere et. al. [135] divided the fingers into two groups: fingers without defects in their structure designated by C1 and fingers with defects, designated by C2. C1 fingers have been observed by many authors and particularly by Press and Arrott [27, 136]. Furthermore, our experimental results showed that the unidirectional spirals, formed under the action of a DC voltage, are with a double twisting of the director [137]. It is well- known that both C1 and C2 fingers can form unidirectional spirals under the action of a DC voltage. Although spirals can be created under the action of AC electric field [138-140], we will not focus our attention on them.

3.7.1. Experimental Results

The greater part of the unidirectional spirals obtained by us have been observed in a mixture of 89 wt. % MBBA – 10 wt. % 5CB – 1 wt. % CC, or in a mixture in which the cholesteric CC was replaced by the cholesteric CN. As noted above, these mixtures have a small positive dielectric anisotropy between 1.5 and 2. CC (cholesteryl chloride) is cholesteric with right winding of the director while CN (cholesteryl nonanoate) is cholesteric with left winding of the director. Our mixtures showed tendency to form bubble domains possessing disclinations in their structure as well as double twist in each finger.

a. Compounds and Sample Preparation

We have used the nematic phase of the following liquid crystals: MBBA, 5CB, and 440 with a nematic phase between 0^0C and 75^0C (the liquid crystal 440 is a mixture of 2/3 BMAOB and 1/3 BHAOB). Additionally, we have used the cholesteric liquid crystals: cholesteric chloride (CC), considered previously and cholesteryl nanoate (CN) shown below:

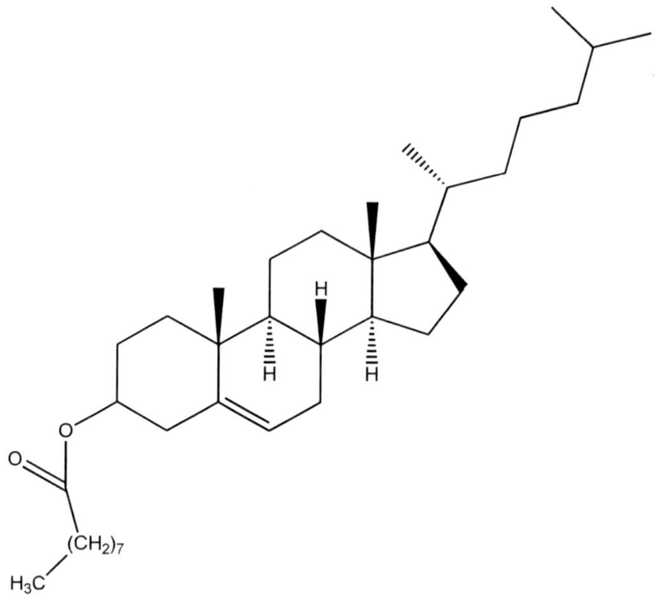

CN: I 92^{0}C Ch 78^{0}C C

On the basis of these liquid crystals, we have prepared the following four (4) liquid crystal mixtures:

1) 89 wt. % MBBA – 10 wt. % 5CB – 1 wt. % CC
2) 88 wt. % MBBA – 10 wt. % 5CB – 2 wt. % CC
3) 89 wt. % MBBA – 10 wt. % 5CB – 1 wt. % CN
4) 89 wt. % 440 – 10 wt. % 5CB – 1 wt. % CC

As noted, the dielectric anisotropy of these mixtures was measured to be between + 1.5 and +2. The cholesterics CC or CN were added to produce either right-handed or left-handed types of the mixtures [141], respectively, while the nematics with negative dielectric anisotropy MBBA or 440 were used to cause either homeotropic (MBBA) or planar (440) orientation of the liquid mixtures. The liquid crystal was confined between two semitransparent and conductive from the one side glass plates prepared by the usual rubbing technique which at appropriate direction of the rubbing and appropriate position of the glass plates relatively each other can ensure either a homogeneous high-tilted (80-85^0) for mixtures containg MBBA or a homogeneous low-tilt (5-10^0) for the mixture which contains the liquid crystal 440. Some of the glass plates were additionally treated by soap with the aid of the simple technique, described above. Non-treated glass plates were studied as well.

b. Experimental Observations

At first the MBBA-5CB-CC mixture was investigated. The non-disturbed by external influences pitch was measured to be around 12 μm in fairly good agreement with the theoretical formula obtained by Arnould and Rondelez [120]:

$$PC = 0{,}12\ \mu m \tag{3.20}$$

In this formula, P is the pitch of the large-pitch cholesteric and C is the concentration of the cholesteric component. It is valid only for small concentration below 5 wt. %, in agreement with our experimental results. The initial state of such a cholesteric mixture with a high tilt at the boundaries, is either Grandjean-like with a screw axis along Z, i.e., normal to the glass plates which usually terminates near the two glass plates either with "scroll" texture due to the conflicting boundary conditions or with a complex confocal texture which contains spiral patterns with both handedness of winding, various dislocations of the cholesteric layers and many disclinations (see above). Let us at this point clarify in accordance with Bouligand [99] that in the right-handed spirals the winding from the centre of the spiral to the outer regions is counter-clockwise while for the left-handed spirals the winding is clockwise. In our experiment, the long- time application of a DC voltage on the "finger-print" texture led to the appearance of unidirectional spirals which change their winding at the change in the direction of the submitted electric field. This is illustrated in Figure 3.32.

A toric domain is seen on the same figure as well. [142, 143]. The spirals themselves posses different numbers of winding and sometimes it is seen only the central part of the spiral called by some authors "bump". The kind of spirals observed in our experiment was different: single or double, their number and their mutual orientation were different depending on the orientation of the director at the boundaries, the pitch of the cholesteric, the duration of the DC voltage applied and the current-voltage characteristics of the cells under study, their current-voltage pre-history, etc [144, 145]. The treatment of the glass plates with soap led to the homogenization of the boundary relief removing many of the defects and stopped the formation of the double spiral patterns.

Figure 3.32. A system of unidirectional spiral patterns formed in a 1 wt. % CC – 89 wt. % MBBA – 10 wt. % 5CB mixture with a thickness of 15 μm, subjected under the action of a DC voltage of 4.2 V; room temperature; crossed nicols; 10 small divisions correspond to 54 μm.

The observation of the spirals was especially complicated for thick cells with a thickness above 40 μm. The spiral patterns were stored for days after switch-off of the voltage especially for the case of thicker cells with a thickness between 20 and 50 μm. The winding of the spirals depends on the direction of the electric field, the value of the thickness of the cell and on the type of the cholesteric used in the mixtures (left-handed or right-handed). All these dependences are schematically shown in Figure 3.33.

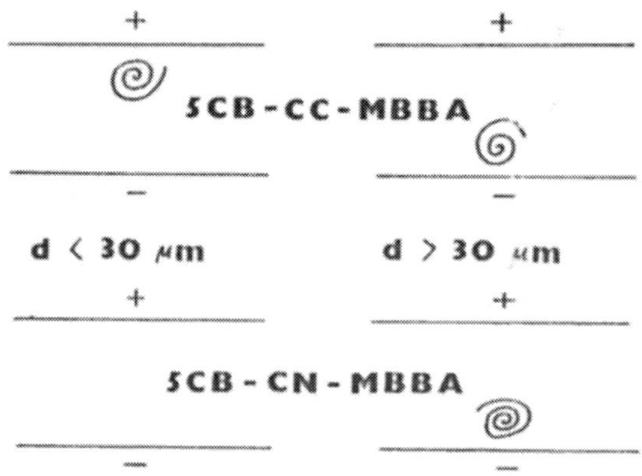

Figure 3.33. The spirals are formed either in the cathode region or at the anode depending on the thickness of the liquid crystal cells, and on the type of the cholesteric (right-handed or left-handed).

However, the application of a DC voltage of 3-5 V across the MBBA-5CB-CC mixture of 30 μm surprisingly led to the formation of right- and left-handed spiral patterns situated near the two Electrodes. The study of the MBBA-5CB-CN mixture, however, did not show such a critical thickness. The spirals were formed only in relatively thick cells of this liquid crystal mixture with a thickness above 30 μm (see Figure 3.34). The increase of the cell thickness to 110 μm removes the formation of the unidirectional spirals for the case of both cholesteric-nematic mixtures. In thick cells we have observed only usual finger-print, confocal or polygonal textures and spirall patterns with both handedness of the winding. Grandjean-like orientation was not observed. Let us point out that unidirectional spirals were observed between 15 and 55 μm. The treatment of the glass plates has no crucial role for the creation of the unidirectional spirals. The homeotropic or nearly-homeotropic orientation of the large-pitch cholesteric in the boundary regions was the only requirement for the formation of the spirals. The value of the surface anchoring, however, was of no importance. In addition, the value of the pitch was of crucial importance for the observation of the unidirectional spirals. For example, at a pitch smaller than 6 μm, spirals did not form at all. We have noted in our experimental investigations the threshold character of the appearance of the unidirectional spirals. This threshold voltage was typically between 3 and 5 V. All experimental findings confirmed that the current-voltage characteristics and the distribution of the electric field inside the cells are of crucial importance for the formation of the spirals. Our measurements indicated that injection of ions existed from the Electrodes, which began at 1 V and continued at bigger voltages as well. We have observed that a small additive of a cholesteric, even of 1 weight percent sharply changed the current-voltage characteristics of the used nematics and led to the appearance of injection for the case of pure nematic MBBA. The appearance of the unidirectional spirals strongly depends on the nature and state of the Electrodes. The long-time duration of the applied high DC voltage (10-15 V) can lead to an irreversible change of the Electrodes which remove completely the spiral patterns. However, when the DC voltage is applied for a short time, the Electrodes might be returned to their initial state by short time application of a high-frequency (above 1 kHz) electric field. The creation of the unidirectional spirals depended on the distribution of the ions not only near the Electrodes but also at distances far away from them.

The cholesteric-nematic mixtures containing the liquid crystal 440 did not show any polar effect possibly due to the strong tendency of the nematic molecules of this liquid crystal to lie in planes, parallel to the Electrodes. The additional treatment of the glass plates with Lecithin, however, permitted the formation of nice spiral patterns like those shown in Figure 3.34 at a DC voltage of U = 4.5 V.

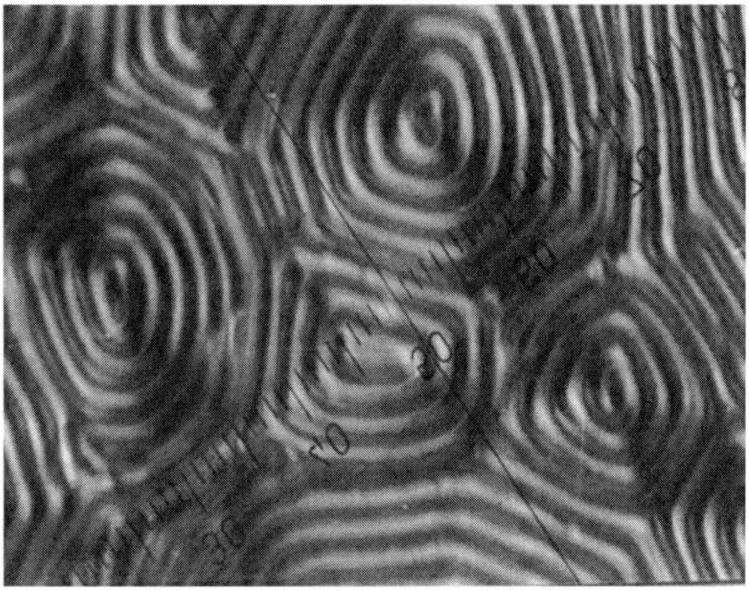

Figure 3.34. Spirals formed in a homeotropic 1 wt. % CC – 89 wt. % 440– 10 wt. % 5CB mixture with a thickness of 15 μm subjected under the action of DC voltage with an amplitude of 4.5 V; room temperature; crossed nicols; 10 small divisions correspond to 54 μm.

The glass plates have been rubbed and then treated by Lecithin in order to ensure homeotropic orientation of the liquid crystal molecules in the boundaries. In thin liquid crystal cells with a thickness below 20 μm, the spiral patterns were very well formed. They were very clearly visible either in an arbitrary position of the two nicols, or even without their use[134]. The change of the focusing in depth of the spiral patterns changes very slightly the image of these (here we should mention that the optical image of the polygons has been drastically dependent on the depth of the focusing as shown by Bouligand [99]).

3.7.2. A Model for Explanation of the Spirals

Three explanations for the creation of the unidirectional spirals under the influence of a DC voltage exist now: our model, suggested by us in 1983 [122], and two theories. The first theory is flexoelectric and has been developed, as noted, by Terentiev and Pikin [123]. The second theory is dynamic one and is developed on the basis of theory of Ginzburg-Landau-de Gennes by Fish, Gil and Gilli [133]. Our model is qualitative and is based on the „gradient flexoeffect". In our opinion, the „gradient flexoeffect" mechanism leads to symmetric deformations of splay-twist-bend in the boundary regions. This mechanism is due to inhomogeneity of the electric field in direction, perpendicular to the plane of the glass plates of the cell. Consequently, we suggested a topological model for the creation of one isolated

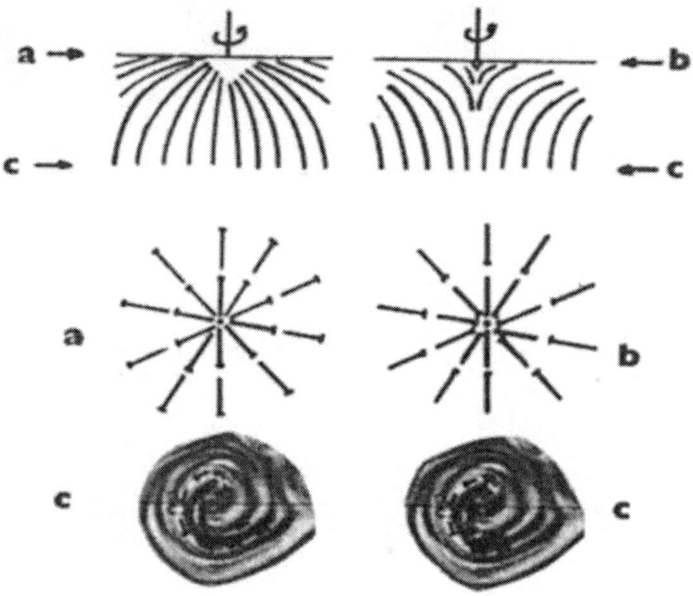

Figure 3.35. A possible topological model for the creation of a single spiral in large-pitch cholesteric containing a point singularity of strength n = +2 (s=+1).

spiral shown in Figure 3.35 for the case of a point disclination of a strength s = +1 (let us point out that our model fixed the spirals and they appeared static under optical observation contrary to the claims of Mitov and Sixou [146]. On the other hand, the same authors have used the idea for the role of the Flexoelectricity in the creation of the spirals [147]. In addition, Gilli and Gil observed that spirals created at a DC voltage do not appear at an AC voltage [148]. The "gradient flexoelectric mechanism" was observed and clarified for the case of nematics [149] (this question will be considered later in chapter 5). This idea, in our opinion, is supported by observation of spiral patterns on cholesteric droplets arising from complex splay-twist-bend deformations of the cholesteric director around point singularities [27, 136]. On the other hand, the strong flexo-electric torques can induce various singularities, even in the case of one unwinded pseudo-nematic. The spiral patterns can be obtained after the symmetric deformation of initially homeotropic orientation which in our case is imposed by the strong dielectric torques acting immediately after the application of the DC voltage. On the other hand, the symmetric splay-twist-bend of the director, leading to the complex curving of the cholesteric layers [150], is caused by the gradient flexoelectric torques due to the non-uniformity of the electric filed along Z direction, perpendicular to the glass plates [149]. It is much more difficult to explain why the spiral patterns change their handedness of winding with increase of the thickness of the cells in the cholesteric-nematic mixtures. We will offer shortly one possible explanation, connected with the injection of ions from the Electrodes (Figure 3.36a).

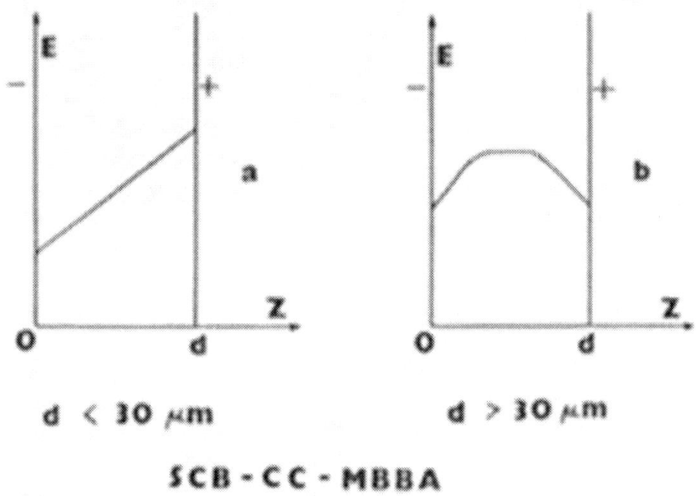

Figure 3.36. A schematic illustration of the electric field distribution in CC-MBBA-5CB mixtures with different thicknesses: a) in thin liquid crystal cells prevail the injection from the cathode and b) in thick liquid crystal cells, the injection from the two Electrodes is important.

In thin liquid crystal cells prevails the injection from the cathode while in the case of thicker cells, the injection from the two Electrodes is more probable (Figure 3,36b) Inversely, the injection from the mixture, containing CN in thin cells is from the anode (Figure 3.37a).

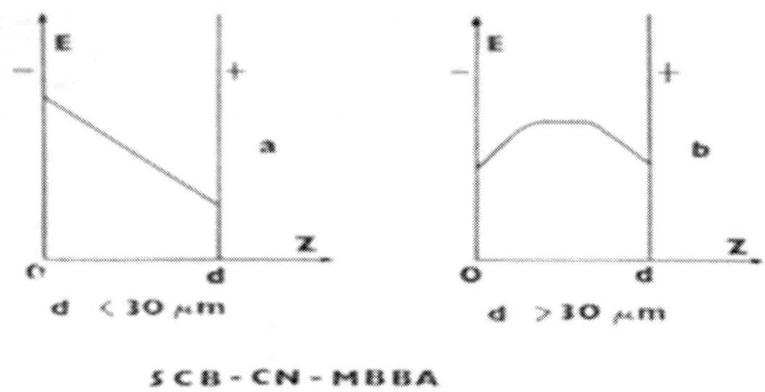

a b

Figure 3.37. A schematic illustration of the electric field distribution in CN-MBBA-5CB mixtures wiyj different thicknesses: a) in thin liquid crystal cells prevail the injection from the anode and b) in thick liquid crystal cells, the injection from the two Electrodes is important.

while for thicker cells, it is probable from the two electodes (Figure 3.37b):

The injection mechanism, which in our opinion governs the spiral formation, can be removed by the application of a voltage with infra-low frequency between 0.1 Hz and 1 Hz which stops the formation of the spirals confirming in this way the importance of this mechanism. In 2005 we noted in an unpublished material that the spirals cannot be formed when there is a 90^0 twist along the axis Z and the cholesteric fingers are deformed under the influence of the boundary (Figure 3.38).

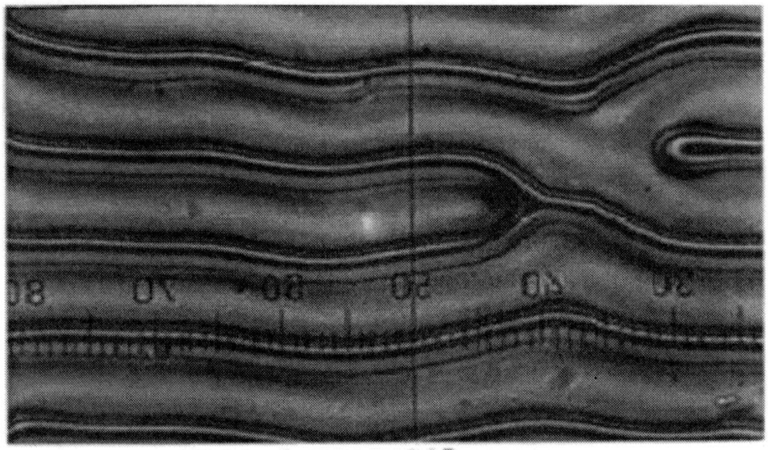

Figure 3.38. A view of the liquid crystal, twisted along Z, perpendicular to the glass plates. It is seen that the cholesteric fingers are deformed under the influence of the two boundaries.

In this case, the application of a DC voltage of 5 V led to the formation of only injection domains, not spirals [151], observable in crossed nicols.

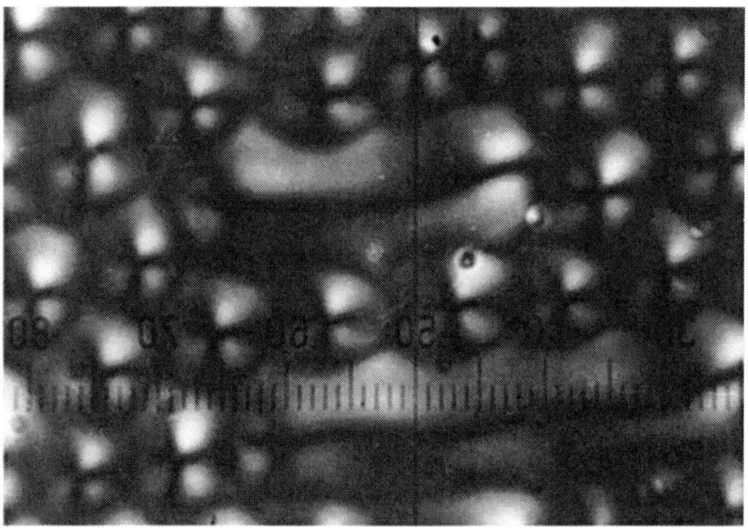

Figure 3.39. The spirals do not appear when a DC voltage of 5 V is applied on the liquid crystal, shown inFigure 3.38. It is seen only cross-like electro-hydrodynamic injection domains of the Felici type. Crossed nicols.

This experimental finding shows that the flexoelectric models are more probable for explanation of the spirals as compared to the dynamic ones. In our opinion, each suggested model, can be connected with a concrete kind of the observed spirals. However, it is of importance to elaborate the theoretical and experimental investigations in this direction.

CONCLUSION

1) Large-pitch cholesteric films are created with strong-strong, strong-weak and weak-weak surface anchoring and a ratio d/p smaller than 0.25, where d is the thickness of the liquid crystal films and p is the pitch of twisting. The electro-optic behavior of these films is investigated and electro-optic transient and hysteretic curves are obtained. It is found that a hysteresis with a larger area is achieved at a weak-weak anchoring and a scanning rate of the voltage from 1 to 3 V/sec.
2) The textures and the electro-optic characteristics of a mixture of nematic liquid crystal-ferroelectric liquid crystal (in particular PCH-7 – DOBAMBC) are observed representing a large-pitch cholesteric with unusual electro-optic and optic properties. They showed intensive colors of the Grandjean texture and very small reaction times of the order of several milliseconds when one realizes electro-optic cholesteric-pseudo-nematic-cholesteric transition.
3) For the first time, unidirectional spirals are observed in large-pitch cholesteric textures "finger-print" embedded into a homeotropic pseudonematic under a DC voltage excitation of the liquid crystal cells. A model is created which explains the formation of spirals in the boundary regions under the influence of a „gradient flexoeffect". The spirals are polar and preserve their kind for a long time (days) after the removal of the voltage.

REFERENCES

[1] De Gennes, P. G. *Sol. St. Commun.* 1968, 6, 163-165.
[2] Meyer, R. B. *Appl. Phys. Lett.* 1968, 12, 281-282.
[3] Leslie, F. M. *Mol. Cryst. Liq. Cryst.* 1970, 12, 57-72.
[4] Kedney, P. J.; Stewart, I. W. *Lett. Math. Phys.* 1994, 31, 261-269.
[5] Regaya, B.; Gasparoux, H.; Prost, J. *Rev. Phys. Appl. (France)* 1972, 7, 83-88.
[6] Dreher, R. *Sol. St. commun.* 1973, 13, 1571-1574.
[7] Malkin, V. M. *Zh. Eksp. Teor. Fiz. (*USSR) 1983, 85, 166-175.
[8] Gotra, Z. Yu.; Mikityuk, Z. M.; Megyei, G. Y.; Zaremba, P. E. *Liq. Cryst.(UK)* 1991, 9, 893-898.
[9] Wysocki, J. J.; Adams, J.; Haas, W. *Phys. Rev. Lett.* 1968, 20, 1024-1025.
[10] Meyer, R. B. *Appl. Phys. Lett.* 1969, 14, 208-209.
[11] Durand, G.; Léger, L.; Rondelez, F.; Veyssie, M. *Phys. Rev. Lett.* 1969, 22, 227-228.
[12] Kwok, S. K.; Liao, Y.*J. Appl. Phys.* 1978, 49, 3970-3974.
[13] Kerlenevich, B.; Coche, A. *Mol. Cryst. Liq. Cryst.* 1983, 97, 103-109.
[14] Kawachi, M.; Kogure, O. *Jpn. J. Appl. Phys.* 1977, 16, 1673-1678.
[15] Kawachi, M.; Kogure, O.; Kato, K. *Jpn. J. Appl. Phys.* 1978, 17, 391-394.
[16] Greubel, W. *Appl. Phys. Lett.* 1974, 25, 5-8.
[17] Watson, P.; Anderson, J. E.; Sergan, V.; Bos, P. J. *Liq. Cryst.(UK)* 1999, 26, 1307-1314.
[18] Watson, P.; Anderson, J. E.; Bos, P. J. *Phys. Rev. E* 2000, 62, 3719-3723.
[19] Anderson, J. E.; Watston, P.; Ernst, T.; Bos, P. J. *Phys. Rev. E,* 2000, 61, 3951-3960.

[20] Yang, D.-K.; Huang, X.-Y.; Zhu, Y-M.. *Ann. Rev. Mater. Sci.* 1997, 27, 117-146.
[21] Helfrich, W. *Appl. Phys. Lett.* 1970, 17, 531-532.
[22] Helfrich, W. *J. Chem. Phys.* 1971, 55, 839-842.
[23] Hurault, J. P. *J. Chem. Phys.* 1973, 59, 2068-2075.
[24] Raynes, E, P. In *Nonemissive Electrooptic Displays*; Kmetz, A. R.; Von Willisen, F. K.; Eds.; Plenum Press, New York, U.S. 1976, pp 26-43.
[25] Raynes, E. P.; In *Liquid crystals of One and Two-Dimensional Order*; Helfrich, W.; Heppke, G.; Eds.; Springer-Verlag, Berlin, Germany 1980, pp 176-184.
[26] Chilaya, G. S. *Kristallografia (Russia) (In Russian)* 2000, 45, 944-960.
[27] Press, M. J.; Arrott, A. S. *Mol. Cryst. Liq. Cryst.* 1976, 37, 81-99.
[28] Stieb, A. E. *J. Phys.(Paris)* 1980, 41, 961-969.
[29] Cladis, P. E.; Kléman, M. *Mol. Cryst. Liq. Cryst.* 1972, 16, 1-20.
[30] Ranganath, G. S. *Mol. Cryst. Liq. Cryst.* 1983, 94, 285-290.
[31] Smalyukh, I. I.; Lavrentovich, O. D.; *Phys. Rev. Lett.* 2003, 90, 085503 (4 pp).
[32] Van Sprang, H. A. *Mol. Cryst. Liq. Cryst.* 1984, 122, 269-283.
[33] Gerber, P. R. *Zeit. Naturforsch. (a)* 1983, 38, 407-414.
[34] Schadt, M.; Gerber, P. *Mol. Cryst. Liq. Cryst.* 1981, 65, 241-264.
[35] Bartolino, R.; Ruffolo, A.; Simoni, F. Scsramuzza, N. *Nuovo Cim. D* 1982, 1, 607-614.
[36] Scheffer, T. J. *Phys. Rev. Lett.* 1972, 28, 593-596.
[37] Hervet, H.; Hurault, J. P.; Rondelez, F. *Phys. Rev. A* 1973, 8, 3056-3064.
[38] Nigemann, E.; Stegemeyer, H. *Liq. Cryst,(UK)* 1989, 5, 739-747.
[39] Ishikawa, I.; Lavrentovich, O. D. *Phys. Rev. E* 2001, 63, 030501 (4 pp).
[40] De Gennes, P. G. *The Physics of Liquid Crtstal;* Clarendon Press, Oxford, England, Table 3.2, 1974.
[41] Chang, R. *Mol. Cryst. Liq. Cryst.* 1974, 28, 1-8.
[42] Leenhouts, F.; van der Woude.; Dekker, A. J. *Phys. Lett. A* 1976, 58, 242-244.
[43] Chen, G.-P.; Takezoe, H.; Fukuda, A. *Liq. Cryst. (UK)* 189, 5, 341-347.
[44] Schell, K. T.; Porter, R. S. *Mol. Cryst. Liq. Cryst.* 1990, 188, 97-104.
[45] Wu, Sh.-T.; Warenghem, M.; Ismaili, M. *Opt. Eng.* 1993, 32, 1775-1780.
[46] Cui, M.; Kelly, J. R. *Mol. Cryst. Liq. Cryst.* 1999, 331, 49-57.
[47] Brugioni, S.; Faetti, S.; Meucci, R, *Opt. Zhurnal (Russia) (In Russian)* 2006, 73, N5, 15-17.
[48] Price, F. P.; Bak, Ch. S. *Mol. Cryst. Liq. Cryst.* 1975, 29, 225-229.
[49] Sartirana, M. L.; Valenti, B.; Bartolino, R. *Mol. Cryst. Liq. Cryst.* 1983, 98, 321-347.
[50] Rondelez, F.; Hulin, J. P. *Sol. St. Commun.* 1972, 10, 1009-1012.
[51] Lin-Hemdel, C. G. *Appl. Phys. Lett.* 1981, 38, 615-617.
[52] Lin-Hendel, C. G. *J. Appl. Phys.* 1982, 53, 916-920.
[53] Gerber, P. R. *Z. Naturforsch. (a)* 1981, 36, 718-726.
[54] Van Sprang, H. A.; Van de Venne, J. L. *J. Appl. Phys.* 1985, 57, 175-179.
[55] Lee, J. C.; Allender, D. W.; Neff, V. D. *Mol. Cryst. Liq. Cryst.* 1992, 210, 11-20.
[56] Lee, J. C.; Allender, D. W. *Jpn. J. Appl. Phys.* 1994, 33, 1407-1411.
[57] Yu. L. J.; Labes, M. M. *Mol. Cryst. Liq. Cryst.* 1974, 28, 423-435.
[58] Goossens, W. J. A. *J. Phys. (Paris)* 1982, 43, 1469-1474.
[59] Barratt, P. J.; Fraser, C. *Proc. Edinburg Math. Soc.* 1983, 26, 319-332.
[60] Mochizuki, A.; Kobayashi, Sh. *Mol. Cryst. Liq. Cryst.* 1993, 225, 89-98.
[61] Schlangen, L. J. M.; Pashai, A.; Cornelissen, H. J. *J. Appl. Phys.* 2000, 87, 3723-3729.

[62] Demus, D.; Richter, L. *Textures of Liquid crystals* ; Verlag Chemie, Weinheim, Germany, 1978.

[63] Shiyanovskii, S. V.; Smalych, L. I.; Lavrentovich, O. D.; In *Liquid crystals: Computer Simulations, Theory and Experiments*; Lavrentovich, O. D. et al.; Eds.; Kluwer Academic Publishers, Amsterdam, The Netherlands, 2001, pp 229-270.

[64] Oswald, P.; Pieranski, P. *Nematic and Cholesteric Liquid crystals* Taylor and Francis, England; 2005, chapter BVII.

[65] Saupe, A. *Mol. Cryst. Liq. Cryst.* 1973, 21, 211-238.

[66] Nawa, N. *Jpn. J. Appl. Phys.* 1990, 29, 560-561.

[67] Bouligand, Y. *J. Phys.(Paris)* 1973, 34, 1011-1020.

[68] Uchiyama, M.; Ishii, Ch. J. *J. Phys. Spc. Jpn.* 1981, 50, 4030-4034.

[69] Hinov, H. P.; Kukleva, E.; Derzhanski, A. I. *Mol. Cryst. Liq. Cryst.* 1983, 98, 109-124.

[70] Kashnow, R. A.; Bugelow, J. E.; Cole, H. S.; Stein, C. R. *Appl. Phys. Lett.* 1973, 23, 290-292.

[71] Hinov. H. P. *Electrooptic and Flexoelectric Properties and Pheniomena In Thermotropic Liquid crystals*; D. Sci. Thesis, Sofia, Bulgaria, 2007, Figures 114a – 114h.

[72] Hinov, H. P.; Mitov, M. D.; Kukleva, E. *Mol. Cryst. Liq. Cryst.* 1986, 136, 281-293.

[73] Rault, J.; Cladis, P. E. *Mol. Cryst. Liq. Cryst.* 1971, 15, 1-13.

[74] Haas, W.; Adams, J.; Dir, G. *Chem. Phys. Lett.* 1972, 14, 95-97.

[75] Berreman, D. W.; Heffner, W. R. *Appl. Phys. Lett.* 1980, 37, 109-111.

[76] Blackburn, W. J. S. *J. Phys D: Appl. Phys.* 1980, 13, 1785-1799.

[77] Gerber, P. R. *Appl. Phys. Lett.* 1984, 44, 932-934.

[78] Blinov, L. M.; Patom S. P. *Liq. Cryst.(UK)* 2009, 36, 1037-1047.

[79] Uchida, T.; Shishido, Ch.; Wada, M. *Electronics and Commun. Jpn.* 1974, 57-C, 103-112.

[80] Brokx, J.; Vertogen, G.; Van Groesen, E. W. C. *Z. Naturforsch. (a)* 1983, 38, 1-9.

[81] Kedney, P. J.; Stewart, I. W. *Cont. Mech, Thermodyn.* 1994, 6, 141-148.

[82] Senyuk, B. I.; Smalykh, I. I.; Lavrentovich, O. D. *Phys. Rev. E* 2006, 74, 011712 (13 pp).

[83] Uemoto, T.; Yoshino, K.; Inuishi, Y. *Jpn. J. Appl. Phys.* 1980, 19, 1467-1472.

[84] Meyer, R. B.; Liébert, L.; Strzelecki, L.; Keller, P. *J. Phys. Lett.(Paris)* 1975, 36, L-69 – L-71,

[85] Pohl, L.; Eidenschink, D.; Krause, G.; Erdman, D. *Phys. Lett. A* 1977, 60, 421-422.

[86] Bradshaw, M. J.; Raynes, E. P. *Mol. Cryst. Liq. Cryst. Lett.* 1971, 72, 35.

[87] Grebyonkin, M. F.; Beresnev, G. A.; Belyaev, V. V. *Mol. Cryst. Liq. Cryst.* 1983, 103, 1-18.

[88] Schadt, Hp.; Baur, G. Meier, G. *J. Chem. Phys,* 1979, 71, 3174-3181.

[89] Schadt, Hp.; Baur, G.; Meier, G. *J. Chem. Phys* 1979, 70, 2770-2774.

[90] Osman, M. A.; Schadt, Hp.; Zeller, H. R. *J. Chem. Phys.* 1983, 78, 906-914.

[91] Scadt, Hp.; Osman, M. A. *J. Chem. Phys.* 1981, 75, 880-885.

[92] Hoffmann, J.; Kuczynski, W.; Malecki, J.; Pavel, J. *Ferroelectrics* 1987, 76, 61-67.

[93] Takezoe, H.; Kondo, K.; Miyasata, K., Abe, S.; Tsuchiya, T.; Fukuda. A.; Kuze, E. *Ferroelectrics* 1984, 58, 55-70.

[94] Kai, Sh.; Takata, M.; Hirakawa, K. *Jpn. J. Appl. Phys.* 1983. 21, 938-941.

[95] Pikin, S. A.; In *Advances in Liquid crystals and Applications*; Bata, L.; Ed.; Pergamon Press, Oxford-Akademiai Kiado, Budapest, Hungary; 1980, 447-467.
[96] Chigrinov, V. G.; Pozhidaev, E. P.; Baikalov, V. A.; Barnik, M. I. *Kristallografiya (USSR)*1989, 34, 406-412.
[97] Abdulhalin, I.; Benguigui, I.; Weil, R. *J. Phys. (Paris)* 1985, 46, 1429-1433.
[98] Galerne, Y.; Poinsot, I.; Schaegis, D. *Appl. Phys. Lett.* 1997, 71, 222-224.
[99] Bouligand, Y. *J. Phys.(Paris)* 1972, 33, 715-736.
[100] Bouligand, Y. *J. Microscopie* 1973, 17, 145-156.
[101] [Bouligand, y. *J. Phys.(Paris)*1973, 34, 603-614.
[102] Bouligand, Y.; Cladis, P. E.; Liebert, L.; Strzelecki, L. *Mol. Cryst. Liq. Cryst.* 1974, 25, 233-252.
[103] Hinov, H. P.; Shonova, N.; Toshev, S. D. *Mol. Cryst. Liq. Cryst.* 1991, 209, 63-77.
[104] Asai, H.; Terasaki, M.; Hasegawa, T.; Kurita, S. *Mol. Cryst. Liq. Cryst.* 1982, 84, 285-307.
[105] Scaramuzza, N.; Bartolino, R.; Barbero, G. *J. Appl. Phyd.* 1982, 53, 8593-8598.
[106] Belyaev, V. V.; Belyaev, S. V.; Chigrinov, V. G.; Grebenkin, M. F.; In *Advances in Liquid crystal Research and Applications* ; Bata, L.; Ed.; Pergamon Press: Oxford-Akademiai Kiado, Budapest, Hungary. 1980, pp 673-691.
[107] Kurik, M.; Mikityuk, Z.; Fechan, A.; Vitak, M.; Nevmerzhitska, O. *Mol. Cryst. Liq. Cryst.* 1997, 293, 1-6.
[108] Pelzl, G.; Eremin, A.; Diele, S.; Kresse, H.; Weissflog, W. *J. Mat. Chem.* 2002, 12, 2591-2593.
[109] Hinov, H. P. *Cholesteric Liquid crystal Material*, Certificat for authorship N 50898.; Reg. N 92452 1990, Sofia, Bulgaria.
[110] Nara, Y.; Kobayashi, S.; Miyaji, A. *J. Appl. Phys.* 1978, 49, 4277-4279.
[111] Van Sprang, H. A.l; Van de Venne, J. L. M. *Proc. Eurodisplay Conf. Paris,* 1984, 207-210.
[112] Clerk, J. F.; Muller, F.; Perrin, A.; Vinet, F. *Digest S.I.D.* 1984, January, 134-136.
[113] Gerber, P. R. *Fourt Display Res. Conf., Eurodisplay;84* 1984, 203-206.
[114] Huang, X.-Y.; Yang, D.-K.; Doane, J. W. *Appl. Phys. Lett.* 1995, 67, 1211-1213.
[115] Sergan, V.; Reznikov, Yu.; Anderson, J.; Watson, P.; Ruth, J.; Bos, P. *Mol. Cryst. Liq. Cryst.* 1999, 330, 95-100.
[116] Watson, P.; Sergan, V.; Anderson, J. E.; Ruth, J.; Bos, P. J. *Liq. Cryst.(UK)* 1999, 26, 731-736.
[117] Sugimura, A.; Kawamura, T. *Jpn. J. Appl. Phys.* 1985, 24, 905-908.
[118] Golovin, A. B.; Shiyanovskii, S. V.; Lavrentovich, O. D. *Appl. Phys. Lett.* 2003, 83, 3864-3866.
[119] Rondelez, F., Thesis UniversitéParis-Sud, May 1973, *Philips Res. Rep. Suppl.* N 2 1974, 1-157.
[120] Arnould-Netillard, H.; Rondelez, F. *Mol. Cryst. Liq. Cryst.* 1974, 26, 11-31.
[121] Kamien, R.; Selinger, J. *J. Phys C: Condens Matter* 2001, 13, R1-R22.
[122] Hinov, H. P.; Kukleva, E. *Mol. Cryst. Liq. Cryst.* 1984, 109, 203-214.
[123] Terentiev, E. M.; Pikin, S. A. *Kristallografiya (USSR)* 1985, 30, 227-235.
[124] Aranson, I. S.; Kramer, L.; Weber, A. *Physica D* 1991, 53, 376-384.
[125] Shagalov, A. G. *Phys. Lett. A* 1992, 165, 412-416.
[126] Sakaguchi, H. *Prog. Theor. Phys.* 1998, 99, 33-43.

[127] Zhang, Sh.; Hu, B.; Zhang, H. *Phys, Rev. E* 2003, 67, 016214 (5 pp).
[128] Aguareles, M.; Chapman, S. J.; Witelski, T. *Phys. Rev. Lett.* 2008, 101, 224101 (4 pp).
[129] Coullet, P.; Plaza, F. *Int. J. Bifurcation & Chaos* 1994, 4, 1173-1182.
[130] Frish, Th. *Physica D* 1995, 84, 601-614.
[131] Shagalov, A. G. *Phys. Lett. A* 1995, 199, 229-232.
[132] Shagalov, A. G. *J. Tech. Phys. (Poland)* 1996, 37, 559-562.
[133] Frisch, T.; Gil, L.; Gilli, J. M. *Phys. Rev. E* 1993, 48, R4199-R4202.
[134] Kondo, K.; Fukuda, A.; Kuze, E. *Jpn. J. Appl. Phys.* 1981, 20, 1779-1785.
[135] Ribière, P.; Oswald. P.; Pirkl, S. J. *J. Physique II France* 1994, 4, 127-143.
[136] Press, M. J.; Arrott, A. S. *J. Phys.(Paris)* 1976, 37, 387-395.
[137] Tuszynski, J. A.; Otwinowski, M.; Dixon, J. M. *J. Phys. Rev. B* 1991, 44, 9201-9213.
[138] Mitov, M.; Sixou, P. *Mol. Cryst. Liq. Cryst.* 1993, 231, 11-28.
[139] Mitov, M. *Action du champ electrique sur des cristaux liquids chiraux en geometrie confine* 1993, D. Sci. Thesis, Univerity of Nice (France) pp 93-104.
[140] Tarasov, O. S.; Krekhov, A. P.; Kramer, L. *Phys. Rev. E* 2003, 68, 031708 (4 pp).
[141] Kazawaguchi, H.; Wada, M. *Jpn. J. Appl. Phys.* 1975, 14, 651-655.
[142] Bouligand, Y. *J. Phys(Paris)* 1974, 35, 215-235.
[143] Baudry, J.; Pirkl, S.; Oswald, P. *Phys. Rev. E* 1999, 59, 5562-5571.
[144] Yoshino, K.; Yamashito, K.; Inuishi, Y. *Jpn. J. Appl. Phys.* 1974, 13, 1471-1472.
[145] Pawlowska, Z. *Acta Phys. Polonica A* 1984, 65, 483-487.
[146] Mitov, M.; Sixou, P. *Mod. Phys. Lett. B* 1995, 9, 929-937.
[147] Mitov, M.; Sixou, P. *Physica B* 1995, 216, 132-140.
[148] Gilli, J. M.; Gil, L. *Liq. Cryst.(UK)* 1994, 17, 1-15.
[149] Derzhanski, A. I. Dr. Sci. Thesis, Sofie, Bulgaria, 1980.
[150] Rault, J.; In *Liquid crystals & Ordered Fluids*; Jonson, J. F.; Porter, R. S.; Eds.: Plenum Press, New York, U.S. 1974,; vol. 2, pp 677-703.
[151] Oswald, P.; Baudry, J.; Pirkl, S. *Phys. Rep.* 2000, 337, 67-96.

Chapter 4

ELECTRO-OPTIC EFFECTS IN NEMATICS: THE FRÉEDERICKSZ TRANSITION IN WEAKLY-ANCHORED AND INITIALLY DEFORMED NEMATIC FILMS

The Fréedericksz transition in nematics has been observed for the first time by Fréedericksz and Zolina [1, 2]. The theory was developed by Zocher [3-5], Oseen [6] and Frank [7] (see also the review [8] and the citations therein). The dynamic behavior of the nematics and cholesterics is described by the equations of Ericksen-Leslie [9-13]. In close relation with the dynamic behavior are the back flow and rotational bulk and surface viscosities.

4.1. FRÉEDERICKSZ TRANSITION: A GENERAL VIEW

Originally, the Fréedericksz transition represents the threshold appearance of one-dimensional deformation in initially oriented planar or homeotropic layer under the action of externally applied either a magnetic or electric field. Saupe [14] was the first who experimentally investigated the Fréedericksz transition in the nematic phase of PAA (para-azoxyanisol) under the action of magnetic field and compared to the theory. Further, Rapini et al., [15] and Deuling [16] have introduced for the first time the local non-homogeneity of the electric field, taking into account the integral character of the Z-component of the dielectric induction D_z which cannot be disregarded at a big value of the dielectric anisotropy. Hareng et al., [17] have considered the influence of the electric field via the deformations of the nematic on the optical birefringence of the incident optical ray (see also the References [14] to [18]). Grüler et al., [19], have determined for the first time the elastic constant of certain nematics with the aid of the Fréedericksz transition. Deuling and Helfrich [20] studied for the first time the Fréedericksz transition in a conductive nematic and found hysteresis in the deformation of one homeotropic nematic under the action of externally applied electric field. Oldano et al., [21] have theoretically considered and experimentally verified the static behavior of the nematic around the Fréedericks threshold. Welford and Sambles [22] considered theoretically electrically-induced deformations for any arbitrary value of the applied electric field. Jakeman and Raynes [23] found theoretically for the first time formulae established on the basic equations for the case of a twisted nematic:

$$\frac{\partial \varphi}{\partial t} = (1/\eta)\left[\varepsilon_a E^2 / 4\pi + K\left(\partial^2 / \partial z^2\right)\varphi\right] \tag{4.1}$$

with solutions:

$$\tau_{rise} = \eta / \left[\left(\varepsilon_a E^2 / 4\pi\right) + Kq^2\right] \tag{4.2}$$

$$\tau_{decay} = \eta / \left[\left(\varepsilon_a E^2 / 4\pi\right) + Kq^2\right] \tag{4.3}$$

where $\eta = \gamma_1$ is the rotational viscosity of the nematic, q is the amplitude of the deformations (this important problem was elaborated in a number of papers [24-32]). For example, Brochard [27] considered for the first time the back flow which depends very strongly on the initial orientation of the nematic: the back flow influences considerably the dynamics of one homeotropic nematic layer and insignificantly the dynamics of one planarly-oriented nematic layer. Schiekel et al., [28] have considered the dynamics of the Fréedericksz transition including also the equations of Ericksen-Leslie. They showed that the rotational viscosity is equal to the following expression:

$$\gamma_1 = (\alpha_3 - \alpha_2)(1 - \alpha) \tag{4.4}$$

where α_2 and α_3 are viscosity coefficients introduced by Leslie [33, 34] and α is the viscosity coefficient of the back flow. The anomalous stiffness of the elastic constants connected with non-uniform attachment of the director was considered in Reference [35]. The relationships between the bulk properties of nematics and their pretilt angle at the boundaries as well as the relationship between the chemical structures of the nematics and their pretilt angle have been developed by Myrvold et al., [36, 37]. Methods for determination of the sign and magnitude of the tilt angle at the boundaries have been considered in References [38] and [39]. In addition, the temperature dependence of the nematic tilt was also considered [40, 41] (it should be mentioned here that the value of the surface tilt is very important for the construction of the liquid crystal displays). Meyerhofer [31] has shown that the threshold character of the Fréedericksz transition is always preserved even a surface tilt in the range of 1-12^0. Gurthler and Casey [42] considered the influence of the surface tilt on the homogeneous alignment of the nematic. Shimoda et al., [43] confirmed the experimental results obtained in Reference [31]. The dynamics of nematic films with a surface tilt under the action of external electric field was considered in [44]. After Leslie [45], Fraser [46] has theoretically considered the Fréedericksz transition in a twisted layer including also the surface tilt. The authors of [47] and [48] considered the influence of the surface tilt on the important case of one deformed twisted layer. The dynamics of the Fréedericksz transition around the threshold at sub-threshold and above-threshold value of applied voltage has been considered in References [49] and [50]. The clearly visible threshold characteristics of pretilted nematic layers were considered also in [51]. In addition, many authors determined the surface tilt, very important for the liquid crystal displays, by various methods [52-66]. The Fréedericksz transition can be observed either under a polarizing microscope in a transmitted polarized light or in a conoscopic regime [67-75]. There are also conoscopic methods which permit to measure the surface tilt of the nematic director provided the orientation of the liquid crystal ensures a good conoscopic image [76-79]. The boundary conditions are very important, especially in the case of weak anchoring of the liquid crystal. Barratt and Jenkins [80] and Jenkins and Barratt [81] have theoretically considered for the first time this important question. The influence of the weak surface anchoring on the Fréedericksz transition has been considered by many authors. We will point out several of them [82-91]. The kind of the surface energy is so important that it can determine even the kind of the Fréedericksz transition: a second order (this is the usual order), or a first order [92-94]. The weak surface coupling is also important for the twisting of the nematic layers [95-110]. The optical properties of nematic layers twisted at smaller angles (less than 90^0) have also been investigated [111-113].

4.2. Surface Orientation and Surface Interactions: A General Review

The surface orientation of the liquid crystal, and particularly of the nematic, is very important for both the research and application point of view [114-134]. It is well- known that the rubbibg of the glass plates confining the liquid crystal with paper, cloth, leather, etc. (the well- known method of Chatelain [135], elaborated in References [136-141]) can orient the nematic or the large-pitch cholesteric, forming in this way monodomains of l;iquid crystals. Berreman [115] introduced for the first time a sinusoidal relief of the rubbed glass plates and considered theoretically its influence on the confining nematic. In this way Berreman pointed to the elastic nature of the orientation of the liquid crystal which prefers to orient itself along the surface grooves. It is clear that this alignment preserves the minimum of the elastic energy of the nematic. This pioneering work was developed by many authors in theoretical or experimental works [142-162]. The rubbing of polymer such as polyvinyl alcohol (*PVA*) [163-169], as well as nylon [170-171] can ensure good alignment not only of nematics or large-pitch cholesterics, but also hardly-orientable smectics (see chapter 1). The kind of the grooves and some of the methods for their creation have also been considered [172-175]. On the other hand, the asymmetric grooves [176-183] caused bistability and increased the angle of observation, a very important parameter for the application of the liquid crystals in the displays [184]. Another widely used method for alignment of the liquid crystals is the evaporation of SiO, SiO_2, SiO_x etc. or other metal oxides under vacuum with a very small thicknesses being in the range of tens or hundreds of angströms, leading to the formation of "inclined columns", relative to the glass planes. Pioneer in this direction was Janning [114, 168] (see also [185-198]). The liquid crystals can also be oriented by inorganic surfaces [117, 199-202]. Complex physicochemical interactions between the liquid crystal molecules and the substances covering the glass plates (in addition to the elastic ones) are very important for the alignment of the liquid crystals via the adhesion, wetting, polar and non-polar surface energy, etc. [203-214]. Here we should mention the interactions between the liquid crystal and various kinds of Lecithin [216-217]. The combination of the rubbing with the use of

Surfactants can be effected in different ways. Usually for producing homeotropic alignment, one uses two ways: first, to add a small quantity of Lecithin (for instance of 0.2 weight percent)[218] and second, rubbing of the glass plates followed by posing of the surfactant on the glass plate (see, for instance [219]). In our opinion, the second method is preferable since the addition of the surfactant, even in a small quantity, changes the phase transitional temperature because it contaminates the liquid crystal. The third method consists of a deposition of the surfactant onto the glass plates followed by their rubbing [220-223]. Feller et al., [224] compared the surface-induced alignment by the method of Seki et al., with the liquid crystal orientation by *SiO* films evaporated under vacuum on the glass plates. They concluded that at the first method [220-223] using rubbed polymers, the short-range molecular interactions are responsible for the alignment of the first monolayer which then aligns the bulk via epitaxi-like interactions. For the case of obliquely deposited SiO films under vacuum evaporation, however, the first monolayer is randomly distributed indicating that the bulk elastic interactions are responsible for the bulk alignment of the liquid crystal. The surface anchoring of liquid crystal molecules on various substrates has been considered in References [225] to [227]. Finally, one should note some of the new techniques, proposed as ion beam etching of surfaces [228], geometrical anchoring at inclined surface of the liquid crystal [229], geometrically-controlled twist transitions in nematic cells [230], alignment of the nematic liquid crystals by inhomogeneous surfaces [231], memory-free conic anchoring of the liquid crystals on a solid substrate [232], multidirectional rubbed liquid crystal cells [233] and the novel approach to modeling nematic liquid crystal cells [234].

4.3. Determination of Weak θ-Polar and φ-Azimuthal Surface Energy by Treatment of the Glass Plates with Usual Soap

It is well-known that Surfactants and in particular, various kinds of soaps can produce tilted nematic layers with a thickness of several tens of the micron [48, 208, 210]. The structure of the deposed soap layer has not hitherto been studied in detail. From our experimental study we only know that very thin soap layers with a thickness of hundreds of angstrom to part of the micron can determine a homogeneous orientation, often tilted or homeotropic orientation of the director with weak anchoring of the liquid crystal. This depends crucially on the thickness of the deposited soap layer. It is accepted, in agreement with our experimental results, that one or several soap molecular layers usually determine strong surface anchoring of the liquid crystal. The weak surface anchoring evidently indicates that our soap films consist of hundreds of monolayers. The eventual deposition of the soap under vacuum evaporation will not be successful since soap grains are formed which cannot orient the liquid crystal [235, 236]. This is also valid if one evaporates other type of material on the Langmuir-Blodgett films [237]. Multi-layers of soap would also orient with difficulty the liquid crystal due to the complexity of the surface, which is usually textured [238]. The dielectric constant of the soap has been measured to be between 2,7 and 2.4 depending on the number of the aliphatic chains (18 versus 22) [239, 240]. Our technology of the glass plate treatment consists of initial rubbing of the glass plates with abrasive paste, cleaning, then rubbing in the same direction with dry soap and second cleaning by a dry cloth in the same direction. This is shown schematically in Figure 4.1.

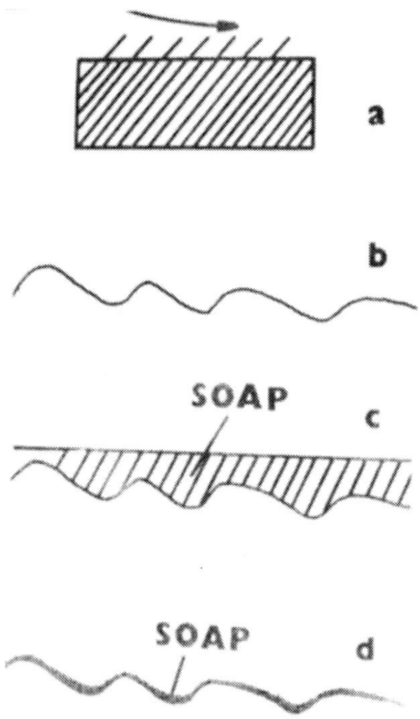

Figure 4.1. A way of the glass plates treatment: a) the rubbing direction can predetermine the sign of the director's surface tilt;b) the kind of the grooves after the first rubbing;c) covering of the glass plate with a thick soap layer;d) the grooves with part of the soap after the second rubbing.

This technology and the pressure exerted on the soap at the rubbing leads to its additional orientation. We believe that our technology of the glass plate treatment with soap and Langmuir-Blodgett technique for deposition of multi-soap layers are too similar [241-243] and in our opinion the orienting capabilities of the two techniques are similar. We assume that the orienting role of the soap consists of two parts: through interaction of the "heads" of the soap molecules with the glass plates or the other substrates and by Van-der-Waals forces of interaction between the liquid crystal and the "tails" of the soap molecules [244-247]. The loosing of the connection between the "heads" of the soap molecules and the glass plate would lead to the formation of two-dimensional nematic structures studied in References from [248] to [250]. The weak anchoring of the liquid crystal layers was achieved by the deposition of Na salt of the fatty acids on the glass plates and the experimental results obtained with such cells were easily reproducible.

4.4. SURFACE ENERGY: FORMS, METHODS OF MEASUREMENTS, BASIC PAPERS

The first scientists who proposed a theoretical formula describing the surface interactions between the nematic director and the bonding surfaces are Rapini and Papoular [251]. The surface potential has been represented as usual sinusoidal function of the boundary polar angle, multiplied by the surface strength constant of the interaction:

$$W_s = C\sin^2\theta /_{z=0,d} \tag{4.5}$$

The relation (4.5) means that the surface energy is positive and that it is minimal for planar orientation of the nematic molecules, resp. of the director. A more general kind of the surface energy has been given by Barratt and Jenkins [80]:

$$W = W\left[(\vec{v}.\vec{n})^2\right] \tag{4.6}$$

The most general form of an surface energy of interaction of the liquid crystal with one arbitrarily deformed surface has been given by Dubois-Violette and Parodi [252]. These authors included in this surface energy terms which depend on the normal derivatives of the director with respect to the coordinates. Jenkins and Barratt [81] considered for the first time the surface orientational moments and the complete boundary conditions. They, however, did not take into account the surface tension and the thermodynamic quantities. Many authors, particularly Barbero and Durand [253], checked the validity of the surface energy, proposed by Rapini and Papoular. These authors and Madhusudana [254] checked the validity of all the forms of the surface energy, proposed before 1984. Barratt and Jenkins elaborated the results of Rapini and Papoular suggesting a more general kind of surface energy. In our opinion, however, Kléman and Pikin proposed for the first time a surface energy which includes not only the polar angle θ, but the azimuthal angle φ as well [255]. These authors included also for the first time a third term representing the "interaction" between the tilt and the azimuthal changes of the director.. Many authors have replaced θ and φ by $(\theta - \theta_0)$ and $\varphi - \varphi_0)$ when the director is tilted at the boundaries. In other words, the elastic minimum in the boundary regions has been translated towards the initial deformations. Pikin[256] gave a more general form of the surface energy which was written as follows:

$$f_s = \frac{1}{2}W_\theta \sin^2(\theta - \theta_0) + \frac{1}{2}W_\varphi \cos^2\theta_0 \sin^2\varphi \tag{4.7}$$

Various kinds of surface energy have been proposed by many authors: [105, 108, 257-263]. The surface energy depends also on the electric double layers and on the flexoeffect when a DC or low-frequency has been used (we will not focus our attention on this problem). Further, the important derivative of the surface energy with respect to the surface deformation angle dW/dθ has been considered by Guyon and Urbach [264] and with the help of this quantity the strength has been measured of the surface interactions without taking into account the form of the surface energy [265, 266]. On the other hand, Barbero and Durand [267, 268] suggested a melting of the nematic due to big surface deformations. There are also many authors who deal with measurement of the surface energy: [118, 119, 123, 134, 269, 270] (see also chapter 7). Another very important problem is the separate measurement of the θ-polar [271-279] and φ-azimuthal [280-301]surface energies. Finally, at a finite anchoring and a strong electric or magnetic field causing the Fréedericksz transition, the external forces can overcome the surface ones and the liquid crystal can be completely oriented [82, 302]. The method for measurement of the surface energy with the aid of a high voltage, well above the threshold, was proposed by Yokoyama and Van Sprang [303]. Various authors dealing with this important problem are cited as follows: [84, 85, 108, 304-310].

4.5. Electro-Optic Dynamic behavior of Weakly-Anchored Nematic Layers with Initial Deformation of Bend-Splay or Twist: Experimental Results

Although there are many theoretical papers devoted to the study of the Fréedericksz transition in nematic cells with weak anchoring, experimental studies are missing. We have obtained in our experiment twist only of 15-20^0 for the case of twisted layers due to the weak φ-azimuthal surface anchoring. The electro-optic transient curves showing the transmittance of the light through the cells curves and the hysteretic curves were compared to those obtained with 90^0 twisted nematic layer achieved by rubbing of the glass plates with the aid of abrasive paste. The threshold in the electrooptic behavior was proved by changes of the conoscopic images (these experimental results are not considered here).

We have investigated a nematic mixture of 90 wt. % MBBA and 10 wt. % 5CB with small positive dielectric anisotropy between 1.5 and 2 (Figure 4.2) [311]:

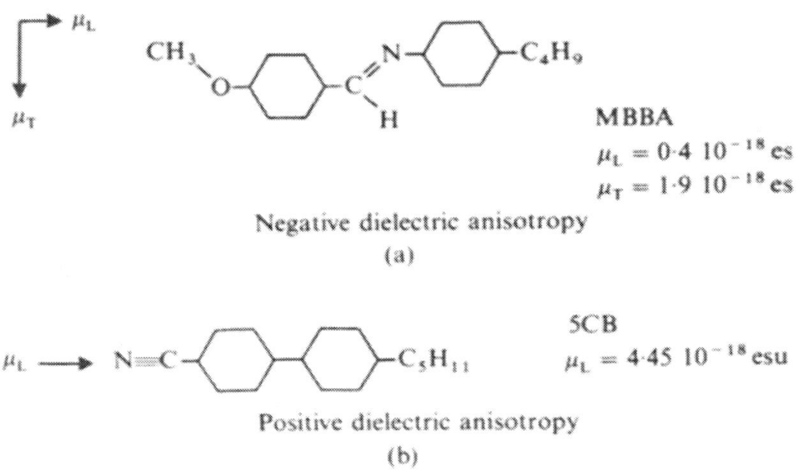

Figure 4.2. Schematic representation of the nematics MBBA and 5CB.

We will show here only the electro-optic and hysteretic curves for the case of weakly anchored and reversely-tilted nematic layers with initial deformation of bend-splay (Figure 4.3).

Figure 4.3. Schematic representation of one deformed homeotropic nematic. The top glass plate was rubbed. The down glass plate was first rubbed and than a soap is posed according to Figure 4.1.

The high initial tilt relative to the glass plate tangent is due to the considerable contribution of the 5CB liquid crystal molecules, which orient themselves homeotropically to the glass plates used by us. We have used polarizing microscope, photomultiplier, X-Y recorder, a special constructed electronic device permitting to obtain a triangular shape of the voltage, i.e., it permits to obtain a linear change of the voltage. This triangular voltage was modulated by a high-frequency voltage in order to avoid the influence of the ions. We observed the changes in a conoscopic regime using a low aperture objective (x10, sin $(\theta_c/2)$ = 0.24), a condenser and a special conoscopic ocular (see Figure 4.4).

The electro-optic curves made by the X-Y recorder are a consequence of the change of the externally applied voltage, which leads to changes in the conoscopic images. The hysteretic curves were recorded at different sweeping rate of the AC voltage up and down. We will show here only the electro-optic transient curves and the hysteretic behavior of the reversely-tilted nematic layer with initial bend-splay deformation and a weak θ-polar anchoring. It is clear that the constructed cells do not show the Fréedericksz transition for planar [312]. [313], homeotropic [314-322] or hybrid (homeo-planar) nematic cells [323].

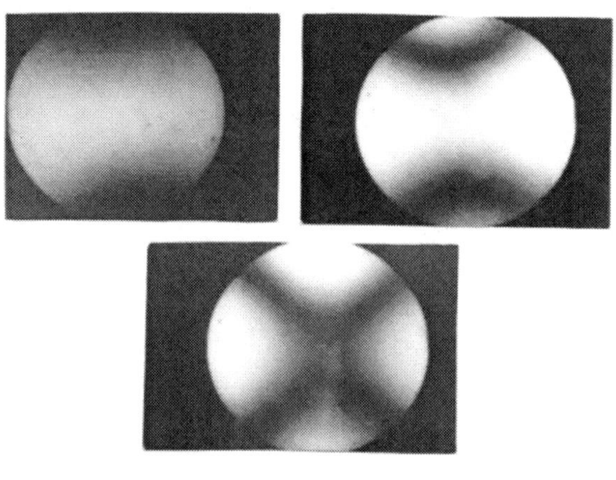

Figure 4.4. Conoscopic picture of the electooptic behaviour of reversely tilted MBBA-5CB mixture with initial deformation of bend- sply and weak θ- polar anchoring. The Fréedericksz transition began with a threshold at 2,5V_{rms} (f=5kHz) and saturated at 25V_{rms}. The conoscopic pictures are shown for the cases 2,5V_{rms}, 5V_{rms} and 10V_{rms}, respectively.

Although the cell under study was with an initial deformation, the Fréedericksz transition started with a threshold of 5 $V_{r.m.s.}$ and a frequency of 5 kHz and at 25 $V_{r.m.s.}$ it saturates (let us recall that the the positive dielectric anisotropy of this mixture is small). The electro-optic transient curves obtained with a X-Y recorder are shown in Figure 4.5 (the curves designated with 1, 2 and 3 correspond to a voltage of 2 $V_{r.m.s.}$, 5$V_{r.m.s.}$ and 10 $V_{r.m.s.}$). These curves show the slow relaxation of the liquid crystal at switch-on and switch-off of the voltage which is in the range of 10 seconds. The dynamic curves [324, 325] clearly showed first, that the soap molecules are not dissolved in the nematic mixture [310, 326, 327] and second, that the relaxation of the nematic director is very slow. Third, there is no development of two-dimensional instability in our films [328]. The weak surface anchoring, the bulk rotational viscosity [329], the inverse flow and the surface viscosity (see chapter 6) contribute to these big relaxation times. This slow relaxation, however, can be used at an appropriate scanning of the liquid crystal cell as is shown in Figure 4.6.

It is seen that the liquid crystal cell has two stable states: "down" and "up". The switching over these two stable states can be realized by the simultaneous application of an impulse with a high amplitude which should switch over the cell from "up" to "down" state. The electro-optic behavior of nematic layers with stronger surface anchoring (only rubbing without utilization of soap) is similar. In this case, however, the threshold and the hysteretic area are significantly smaller and the realization of a bi-stability effect will be hindered. The decrease of the scanning rate of the voltage down to 0.017 V/sec leads to the vanishing of the hysteresis. Since at present a theory of the dynamic hysteresis does not exist, we believe that the slow relaxation of the liquid crystal layers and the fast change of the voltage are the most

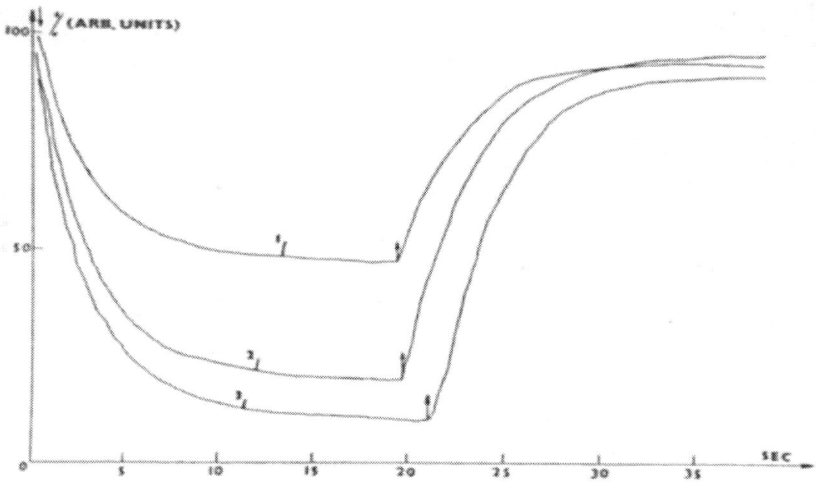

Figure 4.5. Electro-optic transient curves of the 90 wt % MBBA-10 wt % 5CB mixture with deformations of splay-bend. The curves 1, 2 and 3 correspond to 2,5V_{rms}, 5V_{rms} and 10V_{rms}, respectively.

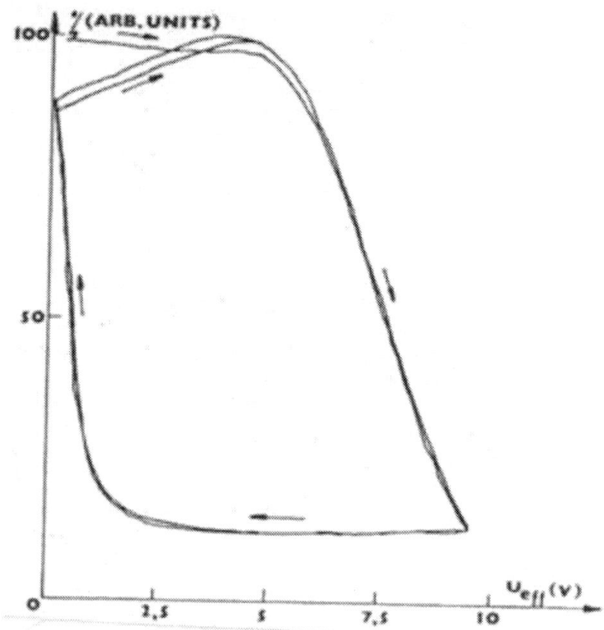

Figure 4.6. The hysteric curves of the same mixture (see Figure 4.5) when the sweeping rate of the AC voltage in the two directions is 0,5 V_{rms}/sec.

important conditions for the existence of such a hysteresis. The strong surface anchoring can reduce the reaction times to milliseconds which evidently removes the possibility for scanning of the cell with a linear change of the voltage, and consequently removes the possibility for appearance of hysteresis. Similar pictures and curves were also observed for the case of twisted nematic layers [311]. However, due to the soap it was not possible to achieve twisting of the nematic by 90^0 (only 20^0) and the realized twisted cells are not convenient for the performance of Schadt-Helfrich effect [330-341]. Finally, let us point out that the hysteretic effects have scarcely been studied [342-347]. However, we believe that our experimental results will stimulate further theoretical and experimental study of this interesting electro-optic effect in nematic liquid crystals.

CONCLUSION

1) For the first time, reversely-tilted and weak anchored nematic films with positive dielectric anisotropy and initial deformation of bend-splay are prepared by using the surfactant – usual soap. One confirms the experimental results and the theoretical calculations for the realization of Fréedericksz transition in electric field that the deformation changes with a threshold and that the relaxations of the director are fairly slow – in the range of seconds. One observes a dynamic hysteresis at fast scanning of the cell with a linearly changing voltage of several volts/sec, which is suitable for application in two-stable nematic cells.
2) The unwinding of nematic films with different twisting and different surface anchoring – weak and middle in the presence of electric field - show also a threshold

character, slow relaxation and a dynamic hysteresis depending on the velocity of the voltage scanning. The hysteretic area, at a given scanning velocity, is larger for cells at weak anchoring.

REFERENCES

[1] Féedericksz, V.; Zolina, V. *Trans. Faraday Soc.* 1933, 29, 910-930.
[2] Fréedericksz, V.; Zwetkoff, V. *Phys. Zeit. Sovwjetunion* 1934, 6, 490-504.
[3] Zocher, H. *Phys. Z.* 1927, 28, 790-796.
[4] Zocher, H. *Trans. Faraday Soc.* 1933, 29, 945-957.
[5] Zocher, H. *Mol. Cryst. Liq. Cryst.* 1969, 7, 165-175.
[6] Oseen, C. W. *Trans. Faraday Soc.* 1933, 29, 683-699.
[7] Frank, F. C. *Discuss. Faraday Soc.* 1958, 25, 19-28.
[8] Ericksen, J. L. In *Advances in Liquid crystals*; Brown, G.; Ed.; 1976; vol. 3, pp 233-298.
[9] Ericksen, J. L. *Arch. Rat. Mech. Anal.* 1960, 4, 231-237.
[10] Ericksen, J. L. *Arch. Rat. Mech. Anal.* 1966, 23, 266-275.
[11] Leslie, F. M. *Quart. J. Mech. Appl. Math.* 1966, 19, 357-370.
[12] Leslie, F. M. *Arch. Rat. Mech. Anal.* 1968, 28, 265-283.
[13] Leslie, F. M. In *IMA: Theory and Applications of Liquid crystals*; Ericksen, J. L.; Kinderlehrer, D.; Eds.; Springer-Verlag, Berlin, Germany, 1987; vol. 5, pp 211-234.
[14] Saupe, A. *Z. Naturforsch. (a)* 1960, 15, 815-822.
[15] Rapini, A.; Papoular, M.; Pincus, Ph. *C. R. Acad. Sc. Paris* 1968, 267, 1230-1233.
[16] Rapini, A. In *Progress in Solid State Chemistry*; McColdin, J.-O.; Somorjar, G.; Eds.; Plenum Press, New York, U.S., 1973; vol. 8, pp 337-401.
[17] Deuling, H. J. *Mol. Cryst. Liq. Cryst.* 1972, 19, 123-131.
[18] Hareng, M.; Leiba, E.; Assouline, G. *Mol. Cryst. Liq. Cryst.* 1972, 361-368.
[19] Akahane, T.; Tako, T.; Masubuchi, S. *Optica Acta* 1979, 26, 943-950.
[20] Deuling, H. J.; Helfrich. W. *Appl. Phys. Lett.* 1974, 25,129-130.
[21] Oldano, C.; Miraldi, E.; Taverna Valabrega, P. *J. Phys.(Paris)*1984, 45, 755-759.
[22] Welford, K. R.; Sambles, J. R. *Mol. Cryst. Liq. Cryst.* 1987, 147, 25-42.
[23] Jakeman, E.; Raynes, E. P. *Phys. Lett. A* 1972, 39, 69-70.
[24] Robert, J.; Labrunie, G.; Borel, J. *Mol. Cryst. Liq. Cryst.* 1973, 23, 197-206.
[25] Pieranski, P.; Brochard, F.; Guyon, E. *J. Phys,(Paris)* 1972, 33, 681-689.
[26] Chigrinov, V. G. *Kristallografiya (USSR)* 1982, 27, 404-430.
[27] Brochard, F. *Mol. Cryst. Liq. Cryst.* 1973, 23, 51-58.
[28] Schiekel, M.; Fahrenschon, K.; Grüler, H. *Appl. Phys.* 1975, 7, 99-105.
[29] Beljaev, V. V.; Blinov, L. M.; Rumjantsev, V. G. *Microelectronica (USSR)* 1976, 5, 544-548.
[30] Baur, G.; Stieb, A.; Meier, G. In *Liquid crystals & Ordered Fluids*; Johnson, J. E.; Porter, R. S.; Eds.; Plenum Press, N.Y., U.S., 1974; vol. 2, pp 645-656.
[31] Meyerhofer, D. *Phys. Lett. A* 1975, 51, 407-408.
[32] Lenzi, E. K.; Barbero, G. *Phys. Rev. E* 2010, 81, 021703 (8 pp).
[33] Leslie, F. M. *Continuum Mech. Thermodyn.* 1992, 4, 167-175.

[34] Ericksen, J. L. *Mol. Cryst. Liq. Cryst.* 1969, 7, 153-164.
[35] Berreman, D. W. *J. Phys. Colloq.(Paris)* 1979, 40, C3-58 – C3-61.
[36] Myrvold, B. O.; Kondo, K.; Oh-Hara, Sh. *Mol. Cryst. Liq. Cryst.* 1994, 239, 211-228.
[37] Myrvold, B. O.; Kondo, K. *Liq. Cryst.(UK)* 1995, 18, 271-286.
[38] Saito, S.; Oyabu, T.; Siracura, Y.; Kamihara, M. *Jpn. J. Appl. Phys.* 1979, 18, 1627-1628.
[39] Opara, T.; Baran, J. W.; Zmiya, J. *Cryst. Res. Technol.* 1988, 23, 1073-1083.
[40] Van Sprang, H. A.; Aartsen, R. G. *Appl. Phys. Lett.* 1983, 42, 669-671.
[41] Van Sprang, H. A.; Aarsten, R. G. *J. Appl. Phys.* 1984, 56, 251-262.
[42] Gurtler, R. W.; Casey, J. W. *Mol. Cryst. Liq. Cryst.* 1976, 35, 275-306.
[43] Shimoda, S.; Mada, H.; Kobayashi, Sh. *Jpn. J. Appl. Phys.* 1978, 17, 1359-1364.
[44] Saito, S.; Yamamoto, H. *Jpn. J. Appl. Phys.* 1978, 17, 395-406.
[45] Leslie, F. M. *Mol. Cryst. Liq. Cryst.* 1970, 12, 57-72.
[46] Fraser, C. *J. Phys. A: Math. Gen.* 1978, 11, 1439-1448.
[47] Scheffer, T. J. *J. Appl. Phys.* 1978, 49, 5835-5842.
[48] Porte, G.; Jadot, J. P. *J. Phys.(Paris)* 1978, 39, 213-224.
[49] Aneva, N.; Petrov, A. G.; Sokerov. S.; Stoylov, S. P. *Mol. Cryst. Liq. Cryst.* 1980, 60, 1-20.
[50] Ou-Yang, Zhong-can.; Xie, Y.-Zh. *Commun. Theor. Phys.(China)* 1983, 2, 1467-1471.
[51] Hauck, G.; Koswig, H. D. *Cryst. Res. Techn.* 1984, 19, 1141-1146.
[52] Baur, G.; Wittwer, V.; Berreman, D. W, *Phys. Lett. A* 1976, 56, 142-144.
[53] Scheffer, T. J.; Nehring, J. *J. Appl. Phys.* 1977, 48, 1783-1792.
[54] Nakano, F.; Isogai, M.; Sato, M. *Jpn. J. Appl. Phys.* 1980, 19, 2013-2014.
[55] Han, K. Y.; Miashita, T.; Uchida, T. *Mol. Cryst. Liq. Cryst.* 1994, 241, 147-157.
[56] Gleeson, J. T. *Liq. Cryst.(UK)* 1996, 20, 453-458.
[57] Toriyama, K.; Ishibashi, T. *Nonemissive Electrooptic Displays*; Kmetz, A. R.; Von Willisen, F. K.; Eds.; Plenum Publishing Corp, New York, U.S., 1976; pp 145-148.
[58] Suzuki, K.; Toriyama, K.; Fukuhara, A. *Appl. Phys. Lett.* 1978, 33, 561-563.
[59] Birecki, H.; Kahn, F. J. In *The Physics and Chemistry of Liquid crystal Devices*; Sprokel, G. J.; Ed.; Plenum Publ. Corp.: New York, U.S., 1980; pp 115-123.
[60] Kosmowski, B. B.; Becker, M. E.; Cremers, R. A.; Mlynski, D. A. *Mol. Cryst. Liq. Cryst. Lett.* 1981, 72, 17-25.
[61] Jain, S. C.; Singth, H. K.; Balakrishnan, K. S.; Chandra, S. *Cryst. Res. Technol.*1987, 22, 1297-1300.
[62] Eidner, K.; Mayer, G.; Rainer, Sch. *Mol. Cryst. Liq. Cryst.* 1987, 152, 447-452.
[63] Han, K.-Y.; Miyashita, T.; Uchida, T. *Jpn. J. Appl. Phys.* 1993, 32, L277-L279.
[64] Han, K.-Y.; Vetter, P.; Uchida, T. *Jpn. J. Appl. Phys.* 1993, 32, L1242-L1244.
[65] Becker, M. E. *J. S.I.D.* 1997, 5/3, 283-287.
[66] Gwag, J. S.; Lee, S. H.; Park, K.-H.; Park, W. S.; Han, K.-Y.; Jhuan, Ch. G. et al. *J. Appl. Phys.* 2003, 93, 4936-4938.
[67] Raynes, E. P.; Rowell, D. K.; Shanks, I. A. *Mol. Cryst. Liq. Cryst. Lett.* 1976, 34, 105-110.
[68] Fahrenschon, K.; Grüler, H.; Schiekel, M. F. *Appl. Phys.* 1976, 11, 67-74.
[69] Fahrenschon. K.; Schiekel, M. F. *J. Electrochem. Soc.* 1977, 124, 953-955.
[70] Johnson, M. R.; Penz, P. A. *IEEE Trans. El. Dev.* 1977, ED-24, 805-807.

[71] Waton, G.; Ferre, A.; Candau, S.; Perbet, J. N.; Hareng, M. *Mol. Cryst. Liq. Cryst.* 1981, 78, 237-249.
[72] Miyoshi, K.; Hashimoto, Y.; Muratomi, K.; Yamashita, M.; Yukinari, T.; Ishiguro, K. *Jpn. J. Appl. Phys.* 1982, 21, L616-L618.
[73] Yamashita, M.; Tako, T.; Ishiguro, K. *Jpn. J. Appl. Phys.* 1988, 27, 469-474.
[74] Yamashita, M. *Jpn. J. Appl. Phys.* 1986, 25, 1-7.
[75] Yamashita, M.; Kurihara, M.; Hisada, E.; Tako, T.; Ishiguro, K. *Jpn. J. Appl. Phys.*1985, 24, 899-904.
[76] Crossland, W. A.; Morrissy, J. H.; Needham, B. *J. Phys. D: Appl. Phys.* 1976, 9, 2001-2014.
[77] Micciancio, S.; Rondelez, F *J. Phys. Lett.(Paris)* 1978, 39, L-5 – L-9.
[78] Galerne, Y. *J. Phys.(Paris)*1978, 39, 1311-1316.
[79] Van Horn, B. L.; Winter, H. H. *Appl. Optics* 2001, 40, 2089-2094.
[80] Barratt, P. J.; Jenkins, J. T. *J. Phys. A: Math. Nucl. Gen.* 1973, 6, 756-769.
[81] Jenkins, J. T.; Barratt, P. J. *Q. Jl. Mech. Appl. Math.* 1974, XXVII, 111-127.
[82] Nehring, J.; Kmetz, A. R.; Scheffer, T. J. *J. Appl. Phys.* 1976, 47, 850-857.
[83] Yamada, H.; Ishii, C. *Mol. Cryst. Liq. Cryst.* 1981, 76, 113-132.
[84] Yang, K. H. *J. Phys.(Paris)* 1983, 44, 1051-1059.
[85] Yang, K. H. *Jpn. J. Appl. Phys.* 1983, 22, 389-393.
[86] Barbero, G.; Madhusudana, N. V.; Durand, G. *J. Phys. Lett.(Paris)* 1984, 45, L-613 – L-619.
[87] Sonnet, A. M.; Virga, E. G. *Phys. Rev. E* 2000, 61, 5401-5406.
[88] Nie, X.; Lu, R.; Xianyu, H.; Wu, Th. X.; Wu., Sh-T. *J. Appl. Phys.* 2007, 101, 103110 (5 pp).
[89] Texeira de Souza, R.; De Jesus, M. M. A.; Dias, J. C.; Evangelista, L. R. *Phys. Rev. E* 2009, 80, 041702 (7 pp).
[90] Dias, J. C.; Yednak, C. A. R.; Lenzi, E. K.; Evangelista, L. R. *Liq. Cryst.(UK)* 2007, 34, 1107-1114.
[91] McKay, G.; Kidney, K. J.; Stewart, I. W. *Mol. Cryst. Liq. Cryst.* 2007, 478, 57-66.
[92] Yang, G.; Shi, J.; Liang, Y. *Liq. Cryst.(UK)* 2000, 27, 875-882.
[93] Shi, J. *Liq. Cryst.(UK)* 2002, 29, 121-125.
[94] Tsui, O. K. C.; Lee, F. K.; Zhang, B.; Sheng, P. *Phys. Rev. E* 2004, 69, 021704 (7 pp).
[95] Chang, T. S.; Greene, P. E.; Loebner, E. E. In *Liquid crystals & Ordered Fluids*; Johnson, J. F.; Porter, R. S.; Eds.; Plenum Press: New York-London, 1973; vol 2., pp 115-123.
[96] Yang, K. H. *J. Appl. Phys.* 1983, 54, 6864-6867.
[97] Becker, M. E.; Nehring, J.; Scheffer, T. J. *J. Appl. Phys.* 1985, 57, 4539-4542.
[98] Strigazzi, A. *J. Phys.(Paris)* 1985, 46, 1507-1512.
[99] Van Sprang, H. A.; Breddels, P. A. *J. Appl. Phys.* 1986, 60, 968-972.
[100] Hirning, R.; Funk, W.; Trebin, H.-R.; Schmidt, M.; Schmiedel, H. *J. Appl. Phys.*1991, 70, 4211-4216.
[101] Derfel, G. *Liq, Cryst.(UK)* 1991, 10, 29-34.
[102] Tariumi, K.; Finkenzeller, U.; Schuler, B. *Jpn. J. Appl. Phys.* 1992, 31, 2829-2836.
[103] Derfel, G. *Liq. Cryst(UK)* 1994, 17, 429-436.
[104] Sugimura, A.; Zhong-can, Ou-Yang, *Mol. Cryst. Liq. Cryst.* 1995, 265, 541-548.
[105] Sugimura, A.; Luckhurst, G. R.; Zhong-can, Ou-Yang *Phys. Rev. E* 1995, 51, 681-689.

[106] Zhou, Y.; He, Zh.; Sato, S. *Jpn. J. Appl. Phys.* 1999, 38, 4857-4858.
[107] Zhao, W.; Wu, Ch.-X.; Iwamoto, M. *Phys. Rev. E* 2002, 65, 031709 (11 pp).
[108] Zakharov, A. V.; Vakulenko, A. A. *Phys. Rev. E* 2005, 72, 021712 (7 pp).
[109] Zhang, S.; An, H.l Guan, R.; Yang, G. *Liq. Cryst.(UK)* 2006, 33, 227-236.
[110] Zhang, Y.-J.; Zhang, Z.-D.; Zhu, L.-Z.; Xuan, L. *Liq. Cryst.(UK)* 2011, 38, 355-359.
[111] Gooch, C. H.;Tarry, H. A. *J. Phys. D:Appl. Phys.* 1975, 8, 1575-1584.
[112] Palmer, S. *Appl. Optics* 1996, 35, 2199-2205.
[113] Palmer, S. *Appl. Optics* 1997, 36, 2094-2100.
[114] Janning, J. L. *Appl. Phys. Lett.* 1972, 21, 173-174.
[115] Berreman, D. W. *Phys. Rev. Lett.* 1972, 28, 1683-1686.
[116] Kahn, F. J.; Taylor, G. N.; Schonhorn, H. *Proc. IEEE* 1973, 61, 823-828.
[117] Cognard, J. *Mol. Cryst. Liq, Cryst.* 1982, suppl. 1, 1-77.
[118] Tomilin, M. G. *Interaction of Liquid crystals with the Bonding Surfaces*; Politechnica, Sankt-Peterburg (Russia)(In Russian), 2001.
[119] Tomilin, M. G. *Opt. Zh (Russia)* 1997, 64, N 5, 58-87.
[120] Chigrinov, V. G.; Belyaev, V. V. *Kristallografiya (USSR)* 1977, 22, 603-607.
[121] Miyano, K. *Jpn. J. Appl. Phys.* 1985, 24, 1379-1388.
[122] Goossens, W. J. A. *Mol. Cryst. Liq. Cryst.* 1985, 124, 305-331.
[123] Luk'anchenko, E. S.; Kozanov, V. A.; Grigos, V. N. *Uspechi Chimii (USSR)* 1985, LIV, 214-238.
[124] Matveenko, V. N.; Kirsanov, E. A. *Uspechi Chimii(USSR)* 1986, LV, 1319-1343.
[125] Blinov, L. M.; Kats, E. I.; Sonin, A. A. *Uspechi Fiz. Nauk(USSR)* 1987, 152, 449-477.
[126] Yokoyama, H. *Mol. Cryst. Liq. Cryst.* 1988, 165, 265-316.
[127] Blinov, L. M.; Kabayenkov, A. Yu.; Sonin, A. A. *Liq. Cryst.(UK)* 1989, 5, 645-661.
[128] Jérôme, B. *Rep. Prog. Phys.* 1991, 54, 391-451.
[129] Schadt, M. *Liq. Cryst.(UK)* 1993, 14, 73-104.
[130] Sonin, A. A. *Surface Physics of Liquid crystals*; Gordon & Breach Sci. Publ.: Langhorne, Luxemburg, 1995.
[131] Barbero, G.; Durand, G. *Liquid crystals in Complex Geometries*; Toledano, P.; Neto, A. M. V; Eds.;World Sc. Publ.: Singapoore, China, 1996.
[132] Plach, H. J.; Reiffenrath, V.; Weber, G. *Proc. SPIE* 1992, 1815, 256-267.
[133] Seeboth, A. *Displays* 1999, 20, 131-136.
[134] Sugimura, A. In *Physical Properties of Liquid crystals: Nematics*; emis Data Reviews; Dunmur, A.; Fukuda, A.; Luckhurst, G. R.; Eds.; INSPEC Publ.: IEE, London, UK 2001; Ser. N 25, pp 493-502.
[135] Chatlain, P. *Compt. Rend. Sc. Acad. Paris* 1941, 213, 875-876.
[136] Verrill, J. F. *J. Phys. E: Sci. Instrum.* 1982, 15, 516-519.
[137] Baber, N.; Strugalski, Z. *Appl. Phys. A* 1984, 33, 209-211.
[138] Pidduck, A. J.; Bryan-Brown, G. P.; Haslam, S. D.; Bannister, R. *Liq. Cryst.(UK)* 1996, 21, 759-763.
[139] Mada, H.; Sonoda, T. *Jpn. J. Appl. Phys.* 1993, 32, L1245-L1247.
[140] Kim, Y, J.; Zhuang, Zh.; Patel, J. S. *Appl. Phys. Lett.* 2000, 77, 513-515.
[141] Rastegar, A.; Skarabot, M.; Blij, B.; Rasing, Th. *J. Appl. Phys.* 2001, 89, 960-964.
[142] Berreman, D. W. *Mol. Cryst. Liq. Cryst.* 1973, 23, 215-231.
[143] Wolff, U.; Greubel, W.; Krüger, H. *Mol. Cryst. Liq. Cryst.* 1973, 23, 187-196.
[144] Sugimura, A.; Yamamoto, N.; Kawamura, T, *Jpn. J. Appl. Phys.* 1981, 20, 1343-1344.

[145] Wen, B.; Rosenblatt, Ch. *J. Appl. Phys.* 2001, 89, 4747-4751.

[146] Strigazzi, A.; Barbero, G.; Miraldi, E.; Ferrero, C. *Mol. Cryst. Liq. Cryst. Lett.* 1983, 82, 345-353.

[147] Lee, E. S.; Vetter, P.; Miyashita, T.; Uchida, T.; Kano, M.; Abe, M.; Sagawara, K. *Jpn. J. Appl. Phys.*1993, 32, L1436-L1438.

[148] Ferrero, C.; Mocella, V.; Sanchez del Rio, M.; Barberi, R. *Mol. Mater.* 1997, 9, 109-122.

[149] Lee, E. S.; Saito, Y.; Uchida, T. *Jpn. J. Appl. Phys.* 1993, 32, L1822-L1825.

[150] Fournier, J.-B.; Galatola, P. *Phys. Rev. E* 1999, 60, 2404-2407.

[151] Hallam, B. I.; Sambles, J. R. *Liq. Cryst.(UK)* 2000, 27, 1207-1211.

[152] Fournier, J.-B.; Galatola, P. *Eur. Phys. J. E* 2000, 2, 59-65.

[153] Kondrat, S.; Poniewierski, A.; Harnau, L. *Liq. Cryst.(UK)* 2005, 32, 95-105.

[154] Batalioto F.; Bechtold, I. H.; Oliveira, E.A.; Evangelista, L. R. *Phys. Rev. E* 2005, 72, 031710 (4 pp).

[155] Fukuda, J.-I.; Yoneya, M.; Yokoyama, H. *Phys. Rev. Lett.* 2007, 98, 187803 (4 pp).

[156] Harnao, L.; Kondrat, S.; Poniewerski, A. *Phys. Rev. E* 2007, 76, 051701 (9 pp).

[157] Fukuda, J.-I.; Yoneya, M.; Yokoyama, H. *Phys. Rev. E* 2008, 77, 030701R (4 pp).

[158] Fukuda, J.-I.; Gwag, J. S.; Yoneya, M.; Yokoyama, H. *Phys. Rev. E* 2008, 77, 011702 (16 pp).

[159] Fukuda, J.-I.; Yoneya, M.; Yokoyama, H. *Phys. Rev. E* 2009, 79, 011705 (13 pp).

[160] Fukuda, J.-I.; Uoneta, M.; Yokoyama, H. *Mol. Cryst. Liq. Cryst.* 2010, 516, 12-25.

[161] Spencer, T. J.; Care, C. M.; Amos, R. M.; Jones, J. C. *Phys. Rev. E* 2010, 82, 021702 (13 pp).

[162] Romero-Enrique, J. M.; Pham. Ch.-T.; Patricio,, P. *Phys. Rev. E* 2010, 82, 011707 (7 pp).

[163] Aoyama, H.; Yamazaki, Y.; Matsuura, N.; Mada, H.; Kobayashi, Sh. *Mol. Cryst. Liq. Cryst. Lett/* 1981, 72, 127-132.

[164] Kutty, T. R. N.' Fischer, A. G. *Mol. Cryst. Liq. Cryst.* 1983, 99, 301-318.

[165] Geary, J. M.; Goodby, J. W.; Kmetz, A.; Patel, J. S. *J. Appl. Phys.* 1987, 62, 4100-4108.

[166] Vetter, P.; Ohmura, Y.; Uchida. T. *Jpn. J. Appl. Phys.* 1993, 32, L1239-L1241.

[167] Faetti, S.; Nobili, M. *Phys. Lett. A* 1996, 217, 133-136.

[168] Kubono, A.; Onoda, H.; Inoue, K.; Tanaka, K.; Akiama, R. *Mol. Cryst. Liq. Cryst.* 2002, 373, 127-141.

[169] Koshida, N.; Kikui, S. *Appl. Phys. Lett.* 1982, 40, 541-542.

[170] Vilfan, M.; Mertelj, A.; Čopič, M. *Phys. Rev. E* 2002, 65, 041712 (7 pp).

[171] Vilfan, M.; Čopič., M. *Phys. Rev. E* 2003, 68, 031704 (5 pp).

[172] Flanders, D. C.; Shaver, D. C.; Smith, H. I. *Appl. Phys. Lett.* 1978, 32, 597-598.

[173] Cheng, J.; Boyd, G. D. *Appl. Phys. Lett.* 1979, 35, 444-446.

[174] Nakamura, M.; Ura, M. *J. Appl. Phys.* 1981, 52, 210-218.

[175] Kawata, Y.; Takatoh, K.; Hasegawa, M.; Sakamoto, M. *Liq, Cryst,(UK)* 1994, 16, 1027-1036.

[176] Bryan-Brown, G. P.; Wood, E. L.; Brown, C. V.; Jones, J. S.; Sage, I. C.; Brett, P. *Proc. Int. Display Workshop* 1997, 261-264.

[177] Brown, C. V.; Bryan-Brown, G. P.; Hui, V. C. *Mol. Cryst. Liq. Cryst.* 1997, 301, 163-168.

[178] Jones, J. C.; Wood, E. L.; Bryam-Brown, G. P.; Hui, V. C. *SID 98 Digest* 1998, 858-861.
[179] Bryan-Brown, G. P.; Brown, C. V.; Sage, I. C.; Hui, V. C. *Nature* 1998, 392/26, 365-367.
[180] Van Haaren, J. *Nature* 1998, 392, 332-333.
[181] Laser Focus World, May 1998, 13.
[182] Brown, C. V.; Towler, M. J.; Hui, V. C.; Bryan-Brown, G. P. *Liq. Cryst.(UK)* 2000, 27, 233-242.
[183] Parry-Jones, L. A.; Edwards, E. G.; Elston, S. J.; Brown, C. V. *Appl. Phys. Lett.* 2003, 82, 1476-1478.
[184] Soref, R. A.; Rafuse, M. J. *J. Appl. Phys.* 1972, 43, 2029-2036.
[185] Guyon, E.; Pieranski, P.; Boix, M. *Lett. Appl. Eng. Sci.* 1973, 1, 19-24.
[186] Meyerhofer, D. *Appl. Phys. Lett.* 1976, 29, 691-692.
[187] Yamashita, M.; Amemiya, Y. *Jpn. J. Appl. Phys.* 1976, 15, 2087-2092.
[188] Goodman, L. A.; McGinn, J. T.; Anderson, Ch. H.; Digeronimo, F. *IEEE Trans. El. Dev.* 1977, ED-24, 795-804.
[189] Pollack, J. M.; Haas, W. E.; Adams, J. E. *J. Appl. Phys.* 1977, 48, 831-833.
[190] Müller, F. *Helv. Phys. Acta* 1978, 51, 483-486.
[191] Hiroshima, K.; Mochizuki, M. *Jpn. J. Appl. Phys.* 1980, 19, 567-568.
[192] Hauck, G. *Phys, St. Solidi (a)* 1980, 62, K123-K125.
[193] Hiroshima, K. *Jpn. J. Appl. Phys.* 1982, 21, L761-L763.
[194] Wilson, T.; Boyd, G. D.; Westerwick, E.-H., Storz, F. G. *Mol. Cryst. Liq. Cryst.* 1983, 94, 359-366.
[195] Hiroshima, K.; Obi, H. *Proc. SID* 1984, 25/4, 287-292.
[196] Sugimura, A.; Kawamura, T. *Jpn. J. Appl. Phys.* 1984, 23, 137-141.
[197] Nobili, M.; Lazzari, C.; Schirone, A.; Faetti, S. *Mol. Cryst. Liq. Cryst.* 1992, 212, 97-106.
[198] Phillips, P. L.; Richardson, R. M.; Zarbakhshi, A.; Halsam, S. D. *Liq. Cryst.(UK)* 1997, 23, 699-708.
[199] Nakamura, M. *J. App. Phys.* 1981, 52, 4561-4567.
[200] Uchida, T.; Ohgawara, M.; Shibata, Y. *Mol. Cryst. Liq. Cryst.* 1983, 98, 149-161.
[201] Sanda, P. N.; Dove, D. B.; Ong, H. L.; Jansen, S, A.; Hoffman, R. *Phys. Rev. A* 1989, 39, 2653-2658.
[202] Uchida, T.; Watanabe, H.; Wada, M. *Jpn. J. Appl. Phys.* 1972, 11, 1559-1565..
[203] Creagh, L. T.; Kmetz, A. R. *Mol. Cryst. Liq. Cryst.* 1973, 24, 59-68.
[204] Kahn, F. J. *Appl. Phys. Lett.* 1973, 22, 386-388.
[205] Haller, I. *Appl. Phys. Lett/* 1974, 24, 349-351.
[206] Proust, J. E.; Ter-Minassian-Saraga, L. *J. Phys. Colloq.(Paris)* 1975, 36, C1-77 - C1-81.
[207] Dubois, J. C.; Gazard, M.; Zann. A. *J. Appl. Phys.* 1976, 47, 1270-1274.
[208] Porte, G. J. Phys.(Paris) 1976, 37, 1245-1252.
[209] Yamashita, M.; Amemiya, Y. *Jpn. J. Appl. Phys.* 1979, 18, 1249-1252.
[210] Naemura, S. *J. Phys. Colloq.(Paris)* 1979, 40, C3-514-C3-518.
[211] Uchida, T.; Ishikawa, K.; Wada, M. *Mol. Cryst. Liq. Cryst.* 1980, 60, 37-52.
[212] Naemura, Sh. *J. Appl. Phys.* 1980, 51, 6149-6159.
[213] Naemura, Sh. *Mol. Cryst. Liq. Cryst.* 1981, 68, 183-198.

[214] Filas, R. W.; Patel, J. S. *Appl. Phys. Lett.* 1987, 50, 1426-1428.
[215] Hiltrop, K.; Stegemeyer, H. *Ber. Bunsenges. Phys. Chem.* 1978,82, 884-889.
[216] Hiltrop, K.; Stegemeyer, H. *Ber. Bunsenges. Phys. Chem.*1981, 85, 582-588.
[217] Hiltrop, K.; Stegemeyer, H. *Ber. Bunsenges. Phys. Chem.*1994, 98, 1209-1213.
[218] Ogawa, F.; Tani, C.; Saito, F. *El. Lett.* 1976, 12, 70-71.
[219] Crawford, G. P.; Ondris-Crawford, R. J.; Žumer, S.; Keast, S.; Neubert, M.; Doane, J. W. *Liq. Cryst.(UK)* 1993, 14, 1573-1585.
[220] Seki, H.; Itoh, Y.; Uchida, T.; Masuda, Y. *Jpn. J. Appl. Phys.* 1990, 29, L2236-L2238.
[221] Seki, H.; Masuda, Y.; Itoh, Y.; Uchida, T. *Mol. Cryst. Liq. Cryst.* 1991, 199, 151-158.
[222] Seki, H.; Itoh, Y.; Uchida, T.; Masuda, Y. *Mol. Cryst. Liq. Cryst.* 1992, 223, 93-98.
[223] Seki, H.; Masuda, Y.; Uchida, T. *Mol. Cryst. Liq. Crys* 1996, 282, 323-329.
[224] Feller, M. B.; Chen, W.; Shen, Y. R. *Phys. Rev. A* 1991, 43, 6778-6792.
[225] Meiboom, S.; Sammon, M. In *The Physics and Chemistry of Liquid crystal Devices*; Sprokel, G. J.; Ed.; Plenum Publ. Corp.: N. Y., U.S. 1980; pp 13-21.
[226] Ohgawara, M.; Uchida, T.; Wada, M. *Mol. Cryst. Liq. Cryst.* 1981, 74, 227-242.
[227] Castellano, J. A. *Mol. Cryst. Liq. Cryst.* 1983, 94, 33-41.
[228] Macdonald, B. F.; Zheng, W.;Cole, R. J.; Underwood, I. *J. Phys. D: Appl. Phys.* 2002, 35, L41-L43.
[229] Lavrentovich, O. D. *Phys. Rev. A* 1992, 46, R722-R725.
[230] Patricio, P.; Telo da Gama, M. M.; Dietrich, S. *Phys. Rev. Lett.* 2002, 88, 245502 (4 pp).
[231] Omg, H. L.; Hurd, A. J.; Meyer, R. B. *J. Appl. Phys.* 1985, 57, 186-192.
[232] Ou Ramdane, O.; Auroy, Ph.; Forgeti, S.; Raspaud, E.; Martinat-Lagarde, Ph.; Dozov, I. *Phys. Rev. Lett.* 2000, 84, 3871-3874.
[233] Huang, Ch.-Y.; Lin, Ch.-h.; Wang, J.-R.; Huang. Ch.-W.; Tsai, M.-Sh.; Fuh, A. Y.-G. *J. Appl. Phys.* 2002, 92, 7231-7235.
[234] Newton, Ch. J. P.; Spiller, T. P. *Mol. Cryst. Liq. Cryst.*2002, 372, 167-178.
[235] Tawata, M.; Morita, Sh., Hattori, Sh. *J. Appl. Phys.* 1987, 62, 3763-3765.
[236] Tawata, M.; Morita, Sh., Hattorim Sh. *Thin Solid Films* 1987, 146, 109-114.
[237] Peterson, I. R.; Russell, G. J.; Earls, J. D.; Sambles, J. R. *Thin Solid Films* 1987, 150, 83-88.
[238] Peterson. I. R.; Earls, J. D.; Girling, I. R.; Russell, G. J. *Mol. Cryst. Liq. Cryst.* 1987, 147, 141-147.
[239] Mann, B.,; Kuhn, H. *J. Appl. Phys.* 1971, 42, 4398-4405.
[240] Careem, M.; Jonscher, A. K. *Phil. Mag.* 1977, 35, 1489-1502.
[241] Pietsch, U.; Barberka, T. A.; Englisch, U.; Stömmer, R. *Thin Solid Films* 1996, 284-285, 387-391.
[242] Pietsch, U.; Barberka, T. A.; Geue, Th.; Stümmer, R. *Nuovo Cimento D* 1997, 19, 393-402.
[243] English, U.; Penacorada, F.; Brehmer, L.; Pietsch, U. *Langmuir* 1999, 15, 1833-1841.
[244] Smith, E. R.; Ninham, B. W. *Physica* 1973, 66, 111-130.
[245] Dubois-Violette, E.; De Genes, P. G. *J. Phys. Lett.(Paris)* 1975, 36, L255-L258.
[246] Dubois-Violette, E.; De Gennes, P. G. *J. Coll. Int. Sci.* 1976, 57, 403-410.
[247] Okano, K.; Murakami, J. *J. Phys. Colloq.(Paris)* 1979, 40, C3-525 – C3-527.
[248] Ryschenkow, G.; Kléman, M. *J. Chem. Phys.* 1976, 64, 404-412.
[249] Kléman, M.; Ryschenkow, G. *J. Chem. Phys.* 1976, 64, 413-416.

[250] Bryan-Brown, G. P.; Wood, E. L.; Sage, I. C. *Nature* 1999, 399/27, 338-340.
[251] Rapini, A.; Papoular, H. *J. Phys. Colloq.(Paris)* 1969, 30, C4-54 –C4-56.
[252] Dubois-Violette, E.; Parodi, O. *J. Phys. Colloq.(Paris)* 1969, 30, C4-57 – C4-64.
[253] Barbero, G.; Durand, G. *J. Phys.(Paris)* 1986, 47, 2129-2134.
[254] Barbero, G.; Madhusudana, N. V.; Durand, G. *Z. Naturforsch. (a)* 1984, 39, 1066-1076.
[255] Kléman, M.; Pikin, S. A. *J. Méc (France)* 1979, 18, 661-672.
[256] Pikin, S. A. *Structural Transformation in Liquid crystals*; Gordon & Breach Sci. Publ.: New York, U.S., 1991.
[257] Shyanovski, S. V.; Smalykh, L. I.; Lavrentovich, O. D. In *Defects in Liquid crystals: Computer Simulations, Theory and Experiments*; Lavrentovich, O. D. et al.; Eds.; Kluwer Academic Publ. Neth., 2001; pp 229-270.
[258] Alexe-Ionescu. A. L.; Barberi, R.; Barbero, G.; Beica, T.; Moldovan, R. *Z. Naturforsch. (a)* 1992, 47, 1235-1240.
[259] Alexe-Ionescu, A.; Barbero, G.; Gabbasova, Z.; Sayko, G.; Zvezdin, A. K. *Phys. Rev. E* 1994, 49, 5354-5358.
[260] Beica, T.; Frunza, S.; Moldovan, R.; Stoenescu, D. N. *Mol. Cryst. Liq. Cryst.* 1997, 301, 39-55.
[261] Shiyanovskii, S. V.; Glushchenko, A.; Reznikov, Yu.; Lavrentovich, O. D.; West, J. L. *Phys. Rev. E* 2000, 62, R1477-R1480.
[262] Zhang. B.; Lee, F. K.; Tsui, O. K. C.; Sheng, P. *Phys. Rev. Lett.* 2003, 91, 215501 (4 pp).
[263] Yang, G.-Ch.; Zhang, Sh.-J.; Han, L.-J.; Guan, R. H. *Liq. Cryst.(UK)* 2004, 31, 1093-1100.
[264] Guyon, E.; Urbach, W. In *Nonemissive Electrooptic Displays*; Kmetz, A. R.; Von Willisen, F. K.; Eds.; Plenum Publ. Corp., New York, U.S., 1976; pp 121-144.
[265] Gleeson. J. T.; Palffy-Muhoray, P. *Liq. Cryst.(UK)* 1989, 5, 663-671.
[266] Blinov, L. M.; Radzhabov, D. Z.; Subachius, D. B.; Yablonsky, S. V. *Pis'ma Zh. E.T.F.(USSR)* 1991, 53, 223-227.
[267] Barbero, G.; Durand, G. *J. Phys. II France* 1991, 1, 651-658.
[268] Barbero, G.; Durand, G. *Mol. Cryst. Liq. Cryst.* 1991, 203, 33-44.
[269] Blinov, L. M. *Electrooptical and Magnetooptical Properties of Liquid crystals* ; Wiley, N.Y., U.S. 1980.
[270] Blinov, L. M.; Chigrinov, G. *Elelectrooptic Effects in Liquid crystal Materials*; Springer-Verlag, Berlin, Germany, 1994.
[271] Haas, G.; Fritsch, M.; Wöhler, H.; Mlynski, D. A. *Liq. Cryst.(UK)* 1989, 5. 673-681.
[272] Seo, D.-Sh.; Muroi, K.-I.; Isogami, T.-R.; Matsuda, H.; Kobayashi, Sh. *Jpn. J. Appl. Phys.* 1992, 31, 2165-2169.
[273] Seo, D.-Sh.; Iimura, Y.; Kobayashi, Sh. *Appl. Phys. Lett.* 1992, 61, 234-236.
[274] Subacius, D.; Pergamenshchik, V. M.; Lavrentovich, O. D. *Appl. Phys. Lett.* 1995, 67, 214-216.
[275] Kedzierski, J.; Raszewski, Z.; Kojdecki, M. A.; Rutkowska, J.; Perkowski, P.; Piecek, W.; Zielinski, J.; Zmija, J.; Lipinska, L. *Proc. SPIE* 1998, 3318, 304-311.
[276] Kedzierski, J.; Miszczyk, E.; Kodjecki., M. A.; Raszewski, Z.; Perkowski, P.; Olifierczuk, M. *Proc. SPIE* 2000, 4147, 190-195.
[277] Yang, F.; Sambles, J. R.; Dong, Y.; Gao, H. *J. Appl. Phys.* 2000, 87, 2726-2735.
[278] Toko, Y.; Akahane, T. *Mol. Cryst. Liq. Cryst.* 2001, 368, 4237-4249.

[279] Kedzierski, J.; Raszewski, Z.; Kojdecki, M. A.; Miszczyk, E.; Lipinska, L.; Bialecka-Florjanczyk, E.; Orzeszko, A. *Proc. SPIE* 2002, 4759, 313-320.
[280] Van Sprang, H. A.; Aartsen, R. G. *Mol. Cryst. Liq. Cryst.* 1985, 123, 355-368.
[281] Faetti, S.; Palleschi, V. *Liq. Cryst.(UK)* 1987, 2, 261-268.
[282] Faetti, S.; Palleschi, V.; Schirone, A. *Nuovo Cimento D* 1988, 10, 1313-1324.
[283] Faetti, S.; Palleschi, V.; Schirone, A. *Liq. Cryst.(UK)* 1991, 10, 95-100.
[284] Nobili, M.; Lazzari, C.; Schirone, A.; Faetti, S. *Mol. Cryst. Liq. Cryst.* 1992, 212, 97-106.
[285] Faetti, S.; Lazzari, C. *J. Appl. Phys.* 1992, 71, 3204-3215.
[286] Lee, E. S.; Uchida, T. *Jpn. J. Appl. Phys.* 1994, 33, L1409-L1422.
[287] Iimura, Y.; Kobayashi, N.; Kobayashi, Sh. *Jpn. J. Appl. Phys.* 1994, 33 L434-L436.
[288] Jiang, M.; Wang, Z.; Sun, R.; Ma, K.; Ma, R.; Huang, X. *Jpn. J. Appl. Phys.* 1994, 33, L1242-L1244.
[289] Sun, R.; Huang, X.; Ma, K.; Wang, Z.; Jiang, M. *Phys. Rev. E* 1994, 50, 1253-1256.
[290] Janossy, I.; Jakli, A. *Mol. Cryst. Liq. Cryst.* 1994, 251, 255-263.
[291] Iimura, Y.; Kobayashi, N.; Kobayashi, Sh. *Jpn. J. Appl. Phys.* 1995, 34, 1935-1936.
[292] Faetti, S.; Nobili, M. *Liq. Cryst.(UK)* 1998, 25, 487-494.
[293] Wood, E. L.; Bradberry, G. W.; Cann, P. S.; Sambles, J. R. *J. Appl. Phys.* 1997, 82, 2483-2487.
[294] Barbri, R.; Dozov, I.; Giocondo, M.; Iovane, M.; Martinot-Lagarde, Ph.; Stoenescu, D.; Tonchev, L. V. *Eur. J. B* 1998, 6, 83-91.
[295] Yang, F.; Sambles, J. R.; Bradberry, G. W. *J. Appl. Phys.* 1999, 85, 728-733.
[296] Chen, L.-Y.; Chen, Sh.-H. *Jpn. J. Appl. Phys.* 2000, 39, L368-L370.
[297] Vilfan, M.; Copic, M. *Mol. Cryst. Liq. Cryst.* 2000, 351, 419-426.
[298] Fonseca, J. G.; Galerne, Y. *Appl. Phys. Lett.* 2001, 79, 2910-2912,
[299] Vilfan, M.; Copic, M. *Mol. Cryst. Liq. Cryst.* 2002, 375, 155-164.
[300] Faetti, S.; Marinelli, P. *Phys. Rev. E* 2005, 72, 051708 (9 pp).
[301] Lee. S. H.; Yoon, T.-H.; Kim, J. Ch. *Phys. Rev. E* 2005, 72, 061705 (6 pp).
[302] Hinov, H. P.; Derzhanski, A. I. *J. Phys. Colloq.(Paris)* 1979, 40, C3-505 – C3-509.
[303] Yokoyama, H.; Van Sprang, H. A. *J. Appl. Phys.* 1985, 57, 4520-4526.
[304] Ong, H. L.; Meyer, R. B.; Hurd, A. J. *J. Appl. Phys.* 1984, 55, 2809-2815.
[305] Yokoyama. H.; Kobayashi, Sh.; Kamei, H. *J. Appl. Phys.* 1987, 61, 4501-4518.
[306] Sugimura, A.; Miyamoto, T.; Tsuji, M.; Kuze, M. *Appl. Phys. Lett.* 1998, 72, 329-331.
[307] Nastishin, Yu. A.; Polak, R. D.; Shiyanovskii, S. V.; Bodnar, V. H.; Lavrentovich, O. D. *J. Appl. Phys.*1999, 86, 4199-4213.
[308] Yang, F. Z.; Cheng, H. F.; Gao, H. J.; Sambles, J. R. *J. Appl. Phys.* 2000, 88, 4553-4557.
[309] Napoli, G. J. *J. Phys. A: Math. Gen.* 2006, 39, 11-31.
[310] Hinov, H. P.; Pavlik, J. I.; Todorova, L.; Marinov, Y. G.; Sridevi, S.; Slaveikova, M. G.; Petrov, A. G.; Rafailov, P. M.; Detlaff-Weglikowska, U. In *New Developments in Liquid crystals and Applications*; Choudhury, P. K.; Ed.; Nova Science Publ., New York, U.S., 2014; pp 151-197.
[311] Hinov, H. P. *Mol. Cryst. Liq. Cryst.* 1984, 111, 57-78.
[312] Klaus, W.; Suzuki, Y.; Tsuchiya, M.; Kamiya, T. *Jpn. J. Appl. Phys.* 1995, 34, 6116-6124.
[313] Kedzierski, J.; Kojdecki, M. A.; Raszewska, Z. *Buletyn WAT* 1996, XLV, 25-45.

[314] Schiekel, M. F.; Fahrenschon, K. *Appl. Phys. Lett.* 1971, 19, 391-393.

[315] Kahn, F. *Appl. Phys. Lett.* 1972, 20, 199-201.

[316] Grüler, H.; Scheffer, T. J.; Meier, G. *Z. Naturfirsch. (a)* 1972, 27, 966-976.

[317] Grüler, H.; Meier, G. *Mol. Cryst. Liq. Cryst.* 1972, 16, 299-310.

[318] Ohtsu, M.; Akahane, T.; Tako, T. *Jpn. J. Appl. Phys.* 1974, 13, 621-629.

[319] Sheng, P. *RCA Review* 1974, 35, 408-432.

[320] Kresse, H.; Pelzl, G. *Phys. Stat. Sol. (a)* 1980, 57, K149-K151.

[321] Wang, H.; Wu, Th.; Zhu, X.; Wu, Sh. T. *J. Appl. Phys.* 2004, 95, 5502-5508.

[322] Mirantsev, L. V. *Fizika Tv. Tela, In Russian* 2007, 49, 370-377.

[323] Matsumoto, Sh.; Kawamoto, M.; Mizunoya, K. *Toshiba Review* 1977, No 108, 32-35.

[324] Clark, M. G.; Leslie, F. M. *Proc. R. Soc. Lond. A* 1978, 361, 463-485.

[325] Lin, B.; Taylor, P. L. *Phys. Lett. A* 1993, 172, 281-284.

[326] Mizutani, W.; Shigeno, M.; Ohmi, M.; Suginoya, M.; Kajimura, K.; Ono, M. *J. Vac. Sci. Technol. B* 1991, 9, 1102-1106.

[327] Wall, G. D.; Cleaver, D. J. *Mol. Phys.* 2003, 101, 1105-1112.

[328] Golovaty, D.; Gross, L. K.; Hariharan, S. L.; Gartland, E. C., Jr. *J. Math. Anal. Appl.* 2001, 255, 391-403.

[329] Zacharov, A. V. *Fiz. Tv. Tela, In Russian* 1998, 40, 1945-1949.

[330] Schadt, M.; Helfrich, W. *Appl. Phys. Lett.* 1971, 18, 127-128.

[331] Baur, G.; Windscheid, F.; Berreman, D. W. *Appl. Phys.* 1975, 8, 101-106.

[332] Leslie, F. M. *Pramana,* 1975, Supplement N1, 41-55.

[333] Goodman, L. A.; Meyerhofer, D.; Divioganni, S. *IEEE Trans. El. Dev.* 1976, ED-23, 1176-1182.

[334] Birecki, H.; Kahn, F. J. *J. Appl. Phys.* 1980, 51, 1950-1954.

[335] Ide, M.; Oh-Ide, T.; Kobayashi, S. *Mol. Cryst. Liq. Cryst.* 1987, 144, 225-234.

[336] Ong, H. L. *J. Appl. Phys.* 1988, 64, 614-628.

[337] Schadt, M. *Mol. Cryst. Liq. Cryst.* 1988, 165, 405-438.

[338] Preist, T. W.; Welford, K. R.; Sambles, J. R. *Liq. Cryst.(UK)* 1989, 4, 103-116.

[339] Sergan, T.; Schneider, T.; Kelly, J.; Lavrentovich, O. D. *Liq. Cryst.(UK)* 2000, 27, 567-672.

[340] Kubono, A.; Kyokane, Y.; Kasajima, Y.; Akiyama, R.; Tanaka, K. *J. Appl. Phys.* 2001, 89, 3554-3559.

[341] Self, R. H.; Please, C. P.; Sluckin, T. J. *Eur. J. Appl. Math.* 2002, 13, 1-23.

[342] Vasil'ev, Yu. V. *Zh. Tekh. Fiz. (USSR)* 1984, 54, 227-232.

[343] Vasil'ev, Yu. V. V. *Vestnik MGU (Moscow), Fizika, Astronomia (USSR)* 1985, 26, N 6, 68-72.

[344] Vasil'ev, Yu. V. *Prikl. Math. Mekh.(USSR)* 1985, 49, 336-340.

[345] Schmiedel, H.; Stanarius, R.; Grigutsch, M.; Hirning, R.; Stelzer, J.; Trebin, H.-R. *J. Appl. Phys.*1993 74, 6053-6057.

[346] Derfel, G. *Mol. Cryst. Liq. Cryst.* 1995, 261, 197-204.

[347] Ishihara, Sh.; Hattori, K.; Sugimura, A. *Mol. Cryst. Liq. Cryst.* 2000, 347, 81-94.

Chapter 5

Flexoelectricity in Nematic Liquid Crystals. Polar Flexoelect. Flexoelectricity in Non-Planar Geometries. The Flexoeffect in Simultaneous Applied DC and AC (< 5 Hz) Electric Fields

5.1. Basic Concepts and Effects

Flexoelectricity denotes curvature-induced electric polarization in the liquid crystal subjected to orientation deformations of the director field **n**. Meyer [1, 2] has introduced for the first time the notation "piezoelectricity" in the liquid crystals in the following way:

$$f_{flexoel.} = - e_{1z}\mathbf{E}\mathrm{div}\mathbf{n} - e_{3x}\mathbf{E}.\mathrm{rot}\mathbf{n}\mathrm{x}\mathbf{n} \qquad (5.1)$$

$f_{flexoel.}$ is the density of the flexoelectric energy, e_{1z} and e_{3x} are the flexoelectric coefficients of splay and bend.

Experiments mostly deal with the converse flexoelectric effect – appearance of torques and orientational deformations in the director field **n**(**r**) due to external electric field **E**(**r**). Meyer has proposed the piezoeffect in the nematics as a dipolar phenomenon. Later, under the suggestion of de Gennes [3], the term "piezoeffect" was replaced by the term "flexoeffect". Initially the flexoelectric polarization of the nematic has been connected with the asymmetric form of the nematic molecules which at deformation display macropolarization [1]. Two of the three basic modes (splay, bend or twist) possess the symmetry of a polar vector. They are the deformations of splay and bend. Thus, these deformations can polarize the initially centrosymmetric structure of the nematic according to the Curie principle:

$$\vec{P}_f = e_{1z}\vec{S} + e_{3x}\vec{B} \qquad (5.2)$$

where the coefficiemts e_{1z} and e_{3x} are the flexoelectric coefficients of splay and bend in Meyer's notation. Flexocoefficients are measured in C/m (in SI system) and an order of magnitude estimation yields e (electron density)/(molecular lemgth) ~ 4.10^{-11}C/m. In the literature, however, usually they are measured in c/g/s. or $dyn^{1/2}$. Equation (5.2) is the constitutive one of the direct flexoelectric effect. Experimental studies of this direct effect in nematics are scarce. Experiments deal mostly with the converse flexoelectric effect emerging with the appearance of torques and orientational deformations in the director field **n**(**r**) due to the external electric field **E**(**r**). In this case the bulk flexoelectric torque can be written in the form $\mathbf{\Gamma}_f = \mathbf{n} \mathrm{x} \mathbf{h}_f$, where $\mathbf{h}_f$ is the bulk flexoelectric component of the molecular field [5, 6]:

$$\vec{h}_f = e_{1z}\left(\vec{E}\nabla.\vec{n} - \nabla\left(\vec{E}.\vec{n}\right)\right) + e_{3x}\left[\vec{E}\times\nabla\times\vec{n} - \nabla\times\left(\vec{E}\times\vec{n}\right)\right] \tag{5.3}$$

This expression can be transformed into an equivalent form:

$$\vec{h}_f = \left(e_{1z} - e_{3x}\right)\vec{h}_{f_1} + \left(e_{1z} + e_{3x}\right)h_{f_2} \tag{5.4}$$

The exact form of this equation (according to Derzhanski and Petrov [5, 6]) shows that the bulk flexodeformations can be observed when there is a gradient in the distribution of either the nematic director or in the external electric field **E**, or both. Effects of the first kind are known as linear (with respect to the electric field) and they depend on the difference of splay and bend flexocoefficients. Those of the second kind are known as gradient effects and they depend on the sum of both flexocoefficients (this is the total flexo-coefficient). It can be easily shown that for planar problems where the director field is confined into a plane, the third term in eq. (5.4) is identically zero. In such a case, with a homogeneous external electric field, the only source of flexo-deformations comes from the surface torque at the liquid crystal-substrate interface $\mathbf{\Gamma}_f = \mathbf{n} \mathrm{x} \mathbf{g}_f$ where $\mathbf{g}_f$ is the flexoelectric surface molecular field [6]:

$$\vec{g}_f = -e_{1z}\left(\vec{E}.\vec{n}\right)\vec{s} + e_{3x}\left[\vec{s}\times\left(\vec{n}\times\vec{E}\right)\right] \tag{5.5}$$

where **s** is the surface normal. A complete solution of a static flexo-electric problem requires the inclusion of the other torques in the torque balance equation such as elastic and dielectric ones in the bulk, elastic, surface anchoring and surface polarization ones at the surface [3, 6, 7]. In the dynamic case, viscous torques are also included, again bulk and surface ones [8]. Finally, in the end of this introduction, one should mention the important in our opinion, papers, devoted to the flexo-electricity [9-17] and to the form of the flexo-electric molecules [18-27], including their constant dipoles of the more extensively studied MBBA and 5CB nematics [28-31]

The theory of Meyer [1] has been created in analogy with the theory of piezoelectricity in solids [32] and is based on the asymmetry of the liquid crystal molecules. This piezoelectric effect (direct or converse) has been called by Meyer “a dipolar piezoelectric effect”. Here we should mention that there are similarities and differences between the piezoelectricity in crystals [32] and the piezoelectricity in nematics [1]. One similarity, for example, is that the splay and bend of both the nematic and the crystal lead to macropolarization. The second similarity is that the “inverse piezoeffect” exists in both the liquid crystals and the crystal. There is, however, one important difference. It concerns the microdislocations in the piezoelectric crystals and the disclinations (lines or point-like) in the nematics. The dislocations in the piezoelectric crystals decrease the net polarization since the piezoelectric effect in the crystals is connected with the crystal lattice whereas around the various disclinations in nematics, one usually observes a very large polarization[33-37]. In the conductive nematic, however, the situation is radically different due to the screening effect of the free ions. According to our information, this important fact has been noted for the first time by de Gennes [3] and Rapini [38]. De Gennes [3] was proposed the term "piezoeffect" to be replace by the term "Flexoelectricity". The next conceptual development in flexoeffect is connected with the term „gradient flexoeffect", proposed originally by Derzhanski [39]. He suggested that the flexoeffect might appear aso in nematics when the electric field is non-homogeneous in the bulk of the liquid crystal due to ions which always exist in a conductive liquid crystal. For the observation of this effect: weak surface anchoring of the liquid crystal, at least at one of the two boundaries, existence of initial tilt of the director and finally, a non-homogeneity of the electric field are required. The combination of these three, very important requirements, leads to the development and observation of gradient flexoelectric deformations at very low DC voltage (below 1 V) according to this idea of Derzhanski [40]. In our early paper [39] we have suggested two types of the non-homogeneous electric field: linear one which demonstrates flexoeffect in the entire cell considering sinusoidal initial deformations of the nematic and hyberbolic one which shows flexodeformations near the boundaries. It is imporiant to stress that in each nematic cell there are impurity ions which separate at the application of a low DC voltage [41-56] (for additional information see [57-66]). Polarized states of hybrid aligned nematic layers are considered in [67], threshold flexoelectric deformations in homeotropic nematic layer in [68]. Furthermore, the ions are generated also by the electrochemical behavior of the nematic and the injection from one or the two Electrodes [57, 58, 69]. Brown and Mottram [70] have shown that the flexoeffect is important also for non-conducting nematic. Later, the gradient flexoeffect was specified by Prost and Marcerou [71]. Separation between the dielectric and flexoelectric deformations was considered in Reference [72] by using iterdigital electrodes [13]. Other authors deal also with the flexoelectric deformations in non-homogeneous electric fields [72-77]. The surface polarization of the liquid crystals, particularly of nematics, is important for asymmetric molecules which show affinity for interaction with the surface. Prost and Pershan [13] are the first authors who bring the idea of surface polarization to the attention of the liquid crystal community. Petrov and Derzhanski [78] have considered this problem in details (see also [79-84]). A basic conclusion is that the surface polarization is more important for homeotropic or highly-tilted nematic layers. Many authors considered the important question about the measurement of the flexoelectric coefficients (this problem will be considered in the last 7th chapter of this work). Although most of the authors considered the flexoelectric effect in a homogeneous electric field (a homogeneous electric field can be created also by a special disposition of the Electrodes [85]), the investigation of the flexoeffect in non-homogeneous electric field is the more interesting case.

5.2. POLAR FLEXOELECTRIC EFFECTS IN NEMATICS

The subtlety and diversity of Flexoelectricity in nematics was shown once again by Helfrich [86] who proposed the concept of a polar flexoelectric effect in the case of interaction of a homogeneous nematic layer with a negative dielectric anisotropy and asymmetric anisotropic interfacial energy (zero and infinity) and a constant electric field applied perpendicularly to the liquid crystal. Later, in connection with the consideration of the second-order elasticity K_{13}, Derzhanski and Hinov [87] elaborated the initial theoretical results obtained by Helfrich. for the case of $K_{13} = 0$ as well. Here, we should point out a number of papers dealing with this problem [6, 88-103]. Helfrich indicated that flexoelectric deformation due to the interaction between applied electric field and the parallel to it flexoelectric polarization, may exist for only one sign of the applied electric field The critical threshold voltage obtained by Helfrich has the form:

$$U_c = \frac{K_{33}}{e_{1z} + e_{3x}} \sim 3\ V$$

where K_{33} is the Frank elastic constant of bend and $e_{1z} + e_{3x}$ is the total flexoelectric coefficient.

Helfrich's concept for a polar effect is elaborated theoretically here for the cases of either negative or positive dielectric anisotropy and homogeneous or homeotropic orientation of the nematic at both signs of the applied electric field (deforming action of the dielectric and flexoelectric torques) and one sign of the applied electric field (orienting action of the dielectric torques and deforming action of the flexoelectric torques) and various strengths of the surface coupling (a unified description of this flexoelectric effect for various experimental geometries – 16 different cases, is presented by Derzhanski et al., [6] as well). The differential equations and the boundary conditions are obtained after minimization of the functional of the liquid crystal feee energy with respect to the deformation angle $\theta(z)$. The solutions demonstrated a deforming action of the linear flexoelectric surface torques as well as an orienting or deforming action of the bulk dielectric torque (see Reference [87], $K_{13}= 0$). The character of the boundary conditions brings about a threshold requirement for these deformations solved numerically for the most interesting cases. The exact solution of the problem allows for the effective preliminary argument of appropriate surface energies of interaction of the liquid crystal molecules with the walls attainable in the experiment. The inclusion of the dielectric anisotropy, as noted by Helfrich, on the one hand changes the polarity by mixed flexo-dielectric deformations.

Case A: A deforming action of the dielectric and flexoelectric torques on initially homeotropic nematic with a negative sign of the dielectric anisotropy or orienting action of the dielectric torques and deforming action of flexoelectrically torques on initially homeotropic nematic with positive dielectric anisotropy (see Figure 5.1 and Figure 5.2).

Case B: A deforming action of the dielectric and flexoelectric torques on initially planar nematic with a positive sign of the dielectric anisotropy or orienting action of the dielectric torques and deforming action of the flexoelectric torques on initially planar nematic with negative sign of the dielectric anisotropy (see Figure 5.1 and Figure 5.2).

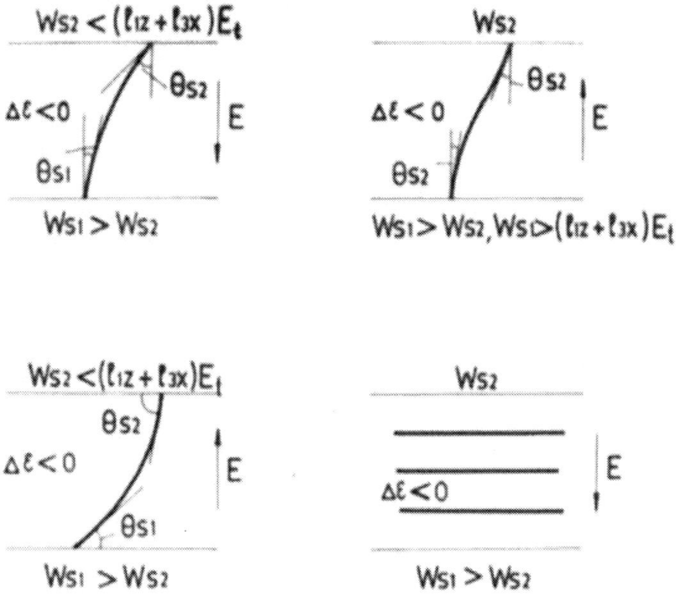

Figure 5.1. Polar flexoelectric deformations in nematic layers with a negative dielectric anisotropy at both directions of the electric field.

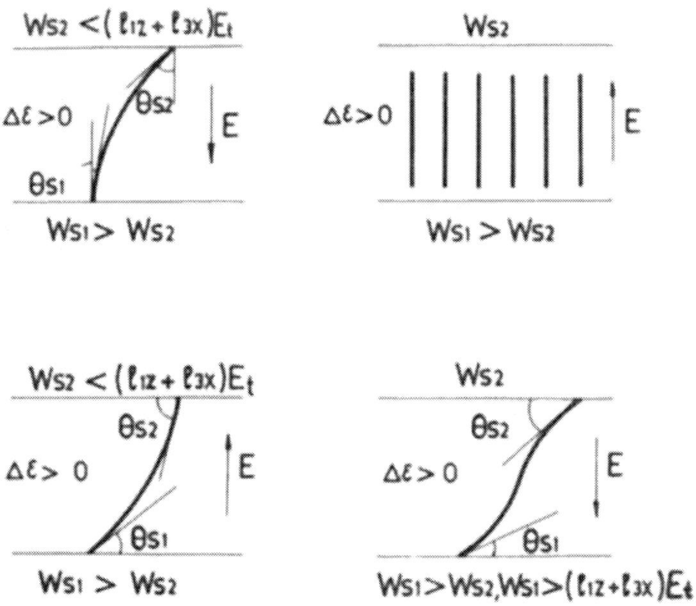

Figure 5.2. Polar flexoelectric deformations in nematic layers with a positive dielectric anisotropy at both directions of the electric field.

We should clarify that first, the flexoelectric torques are with a surface character and depends strongly on the relation between the flexoelectric surface energy and the surface elastic energy and second, that all the deformations (with a maximum inside the nematic cells or with a monotone character) start with a threshold for any direction of the electric field. It follows from here that the operating voltage of this effect can change with the change of the dielectric anisotropy.

The interaction between a thin nematic homeotropic monocrystal with negative or positive dielectric anisotropy and flexoelectric properties as well as finite anisotropic elastic energy of interaction of the interface director with the cell walls W_{s1} and W_{s2}, and a constant electric field, applied normal to the initial orientation of the nematic director, is considered theoretically in this section (the conductivity effects were disregarded).

The free energy of this nematic layer written in invariant vector form following Barratt and Jenkins [104] as well as Meyer [1] reads:

$$\begin{aligned} H_E = \int_V dv \Big\{ (1/2)\left[K_{11}(div\ n)^2 + K_{22}(n.rot\,n)^2 + K_{33}(n\times rot\,n)^2\right] - e_{1z}\,E.n\,div\ n - e_{3x}\,E.rot\,n\times n - \\ -(1/8\pi)\left[\varepsilon_{\parallel}(E.n)^2 + \varepsilon_{\perp}(E\times n)^2\right]\Big\} + W_{s_1}(v_1.n)^2 + W_{s_2}(v_2.n)^2 \end{aligned} \quad (5.6)$$

It is a sum of the elastic energy, the flexoelectric energy and the approximate dielectric energy, accurate to a large extent for nematics with a small dielectric anisotropy ($\Delta\varepsilon/\varepsilon_{\perp} < 1$). The introduction of a rectangular coordinate system OXYZ is most convenient for the problem thus set, where the X axis coincides with the lower interface surface, Z is the normal and Y brings the system to the right orientation. The constant electric field **E**, applied along + Z, is denoted by ↑E and along –Z by ↓E (the x- and y- components of the electric field are zero). The anisotropic interfacial energies ensure that the preferred orientation of the liquid crystal is homeotropic. Substituting the components of the vector field $n_x = \cos\theta$ and $n_z = \sin\theta$ and $E_z = E$ in the expression for the free energy (5.6) transforms the latter into a functional of deformation angle θ(z). Varying this functional with respect to θ(z) gives the deformational equation, describing the deformations in the layer:

$$f(\theta)\left(\frac{d\theta}{dz}\right)^2 + \frac{|\Delta\varepsilon|}{4\pi}E^2\sin^2\theta = C$$

$$f(\theta) = K_{33}\cos^2\theta + K_{11}\sin^2\theta \quad (5.7)$$

and the boundary conditions for the values of θ(z) at the two interfacial surfaces:

$$f(\theta)\frac{d\theta}{dz} + \left[(e_{1z} + e_{3x})E \pm W_{s_{2,1}}\right]\sin\theta\cos\theta = 0, \qquad z = d,0 \quad (5.8)$$

where $d\theta/dz \neq 0$, d is the thickness of the liquid crystal layer.

The following important conclusions can be drawn from equation (5.8): the value of the derivative of the deformation angle at the surfaces depends also on the flexoelectric energy. The sign of the coefficient, multiplying sinθcosθ is of great importance. It is evident that it is determined from the relation between the flexoelectric and the anisotropic interfacial energy.

This sign at the two surfaces dictates the sign of dθ/dz with respect to that of θ(z) at the interfaces. Two types of deformation may be realized with a special selection of the interfacial energies. The monotone deformation is determined by the expression:

$$z = \operatorname{sgn}\theta\left(\frac{|\Delta\varepsilon|}{4\pi}E^2\right)^{-1/2}\int_{\theta_{s1}}^{\theta}\left(\frac{f(\theta)}{k^2-\sin^2\theta}\right)^{1/2}d\theta$$

where

$$\kappa^2 = \frac{f(\theta)\left(\dfrac{d\theta}{dz}\right)^2}{(|\Delta\varepsilon|/4\pi)E^2}+\sin^2\theta, \qquad z=0,d\,|\theta_{s_1}|<(\pi/2); \qquad |\theta_{s_2}|<(\pi/2),$$
$$\frac{d\theta}{dz^{\downarrow E}} = \frac{(e_{1z}+e_{3x})E-W_{s_2}}{f(\theta)}\sin\theta\cos\theta, \qquad z=d$$
$$\frac{d\theta}{dz^{\downarrow E}} = \frac{(e_{1z}+e_{3x})E+W_{s_1}}{f(\theta)}\sin\theta\cos\theta, \qquad z=0 \tag{5.9}$$

The interfacial angles are smaller than π/2 because of the unidirectional action of the bulk and surface torques. This solution requires a small interfacial energy at the upper wall – weak anchoring of the liquid crystal molecules with this wall.

The deformation with a maximum in the liquid crystal layer is determined by the expression:

$$z = \begin{cases} \operatorname{sgn}\theta\left[(|\Delta\varepsilon|/4\pi)E^2\right]^{-1/2}\displaystyle\int_{\theta_{s1}}^{\theta}\left(\frac{f(\theta)}{\sin^2\theta_m-\sin^2\theta}\right)^{1/2}d\theta, \dfrac{d}{dz}\theta^2>0 \\ \operatorname{sgn}\theta\left[(|\Delta\varepsilon|/4\pi)E^2\right]^{-1/2}\left\{\displaystyle\int_{\theta_{s1}}^{\theta_m}\left(\frac{f(\theta)}{\sin^2\theta_m-\sin^2\theta}\right)^{1/2}d\theta-\int_{\theta_m}^{\theta}\left(\frac{f(\theta)}{\sin^2\theta_m-\sin^2\theta}\right)^{1/2}d\theta\right\}, \\ \dfrac{d}{dz}\theta^2<0 \end{cases} \tag{5.10}$$

where

$$\sin^2\theta_m = \frac{f(\theta)\left(\dfrac{d\theta}{dz}\right)^2}{(|\Delta\varepsilon|/4\pi)E^2}+\sin^2\theta, \qquad z=0,d$$
$$\frac{d\theta}{dz_{\uparrow E}} = \frac{-(e_{1z}+e_{3x})E-W_{s_2}}{f(\theta)}\sin\theta\cos\theta, \qquad z=d$$

$$\frac{d\theta}{dz_{\uparrow E}} = \frac{-(e_{1z}+e_{3x})E + W_{s_1}}{f(\theta)}\sin\theta\cos\theta, \qquad z=0$$

$$|\theta_{s_1}| < (\pi/2), \qquad |\theta_{s_2}| < (\pi/2), \qquad W_{s_1} > (e_{1z}+e_{3x})E_t$$

The threshold condition for the monotone deformation is represented by the following formula:

$$\pi\frac{U_1}{U_0} = arctg\, X_{s_1\downarrow E} - arctg\, X_{s_2\downarrow E}$$

$$X_{s_1\downarrow E} = \frac{(e_{1z}+e_{3x})U_0 + W_{s_1}d(U_0/U_1)}{\pi K_{33}}$$

$$X_{s_2\downarrow E} = \frac{(e_{1z}+e_{3x})U_0 - W_{s_2}d\,(U_0/U_1)}{\pi K_{33}} \qquad (5.11)$$

The threshold condition for the deformation with a maximum in the liquid crystal layer is represented by the following formula:

$$\pi\frac{U_1}{U_0} = arctg\, X_{s_1\uparrow E} + arctg\, X_{s_2\uparrow E}$$

$$X_{s_2\uparrow E} = \frac{(e_{1z}+e_{3x})U_0 + W_{s_2}d(U_0/U_1)}{\pi K_{33}}$$

$$X_{s_1\uparrow E} = \frac{-(e_{1z}+e_{3x})U_0 + W_{s_1}d\,(U_0/U_1)}{2\pi K_{33}} \qquad (5.12)$$

The threshold conditions (5.11) and (5.12) can be solved numerically for the following cases: infinite anisotropic interfacial elastic energy of the lower limiting surface (in practice $W_{s1}d = 10^{-4}$ erg/cm and finite anisotropic interfacial energy W_{s2} for MBBA at room temperature ($T = 21^0$ C) and the following values of the total flexoelectric coefficient: 10^{-4}, $3x10^{-4}$, $6x10^{-4}$ in cgs units, U_0 is the voltage threshold for fixed vanishing boundary condition for the homeotropic layer. The numerical calculations for the case of MBBA is shown in the curves illustrated in the figures 5.5-5.5.

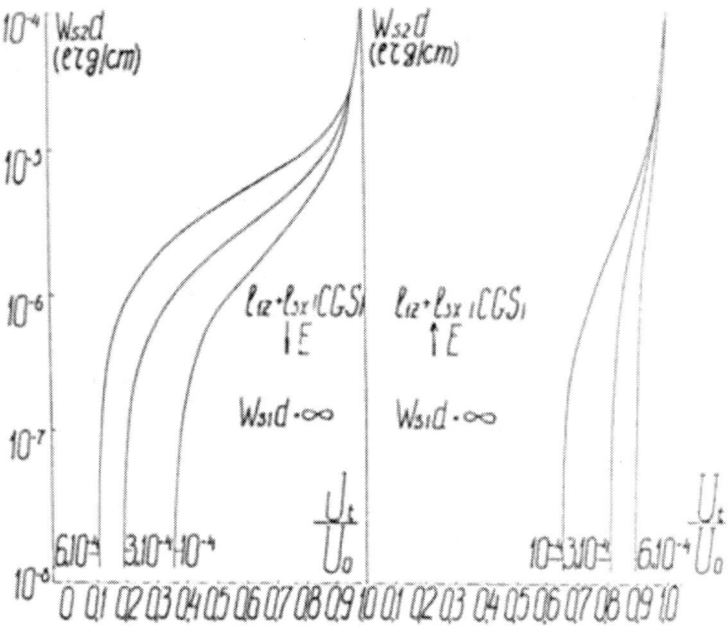

Figure 5.3. Polar theoretical curves at a voltage threshold for a homeotropic MBBA layer and some values of $W_{s2}d(W_{s1}d=\infty)$.

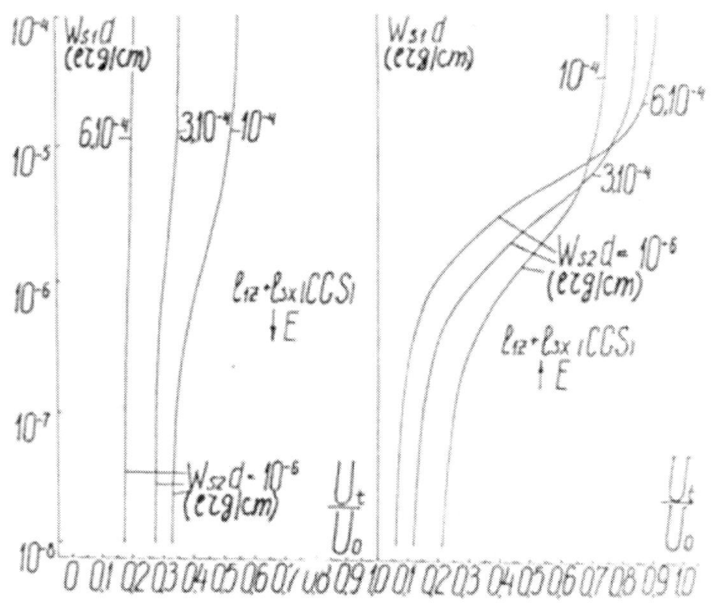

Figure 5.4. Polar theoretical curves at a voltage threshold for a homeotropic MBBA layer and some values of $W_{s1}d$ and a fixed value of $W_{s2}d$.

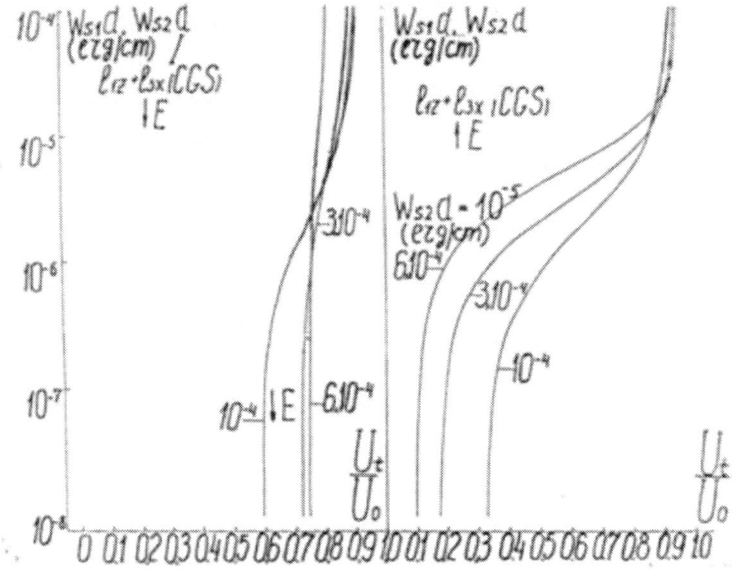

Figure 5.5. Polar theoretical curves at a voltage threshold for a homeotropic MBBA layer for some fixed values of $W_{s2}d$.

In a similar manner the other important case can be solved concerning a planar nematic layer with negative dielectric anisotropy and an electric field applied normally to the unperturbed nematic director, i.e., at stabilizing action of the applied electric field. The surface energies assigned to the two boundaries should be strong and weak. The sign of the total flexocoefficient and the sign of the applied electric field would determine the strength of the polar flxodelectric deformations.

The deformational equation, describing the deformations in the layer determines only a monotone deformation:

$$f(\theta)\left(\frac{d\theta}{dz}\right)^2 - \frac{|\Delta\varepsilon|}{4\pi}E^2\sin^2\theta = C$$

$$f(\theta) = K_{11}\cos^2\theta + K_{33}\sin^2\theta \qquad (5.13)$$

$$z = \operatorname{sgn}\theta\left(\frac{|\Delta\varepsilon|}{4\pi}E^2\right)^{-1/2}\int_{\theta_{s1}}^{\theta}\left[\frac{f(\theta)}{h+\sin^2\theta}\right]^{1/2}d\theta$$

$$h = \frac{f(\theta)\left(\dfrac{d\theta}{dz}\right)^2}{(|\Delta\varepsilon|/4\pi)E^2} - \sin^2\theta, \qquad z = 0, d$$

$$\frac{d\theta}{dz_{\uparrow E}} = \frac{(e_{1z}+e_{3x})E - W_{s_2}}{f(\theta)} + \sin\theta\cos\theta, \qquad z = d$$

$$\frac{d\theta}{dz_{\uparrow E}} = \frac{(e_{1z}+e_{3x})E + W_{s_1}}{f(\theta)} + \sin\theta\cos\theta, \qquad z=0$$
$$(e_{1z}+e_{3x})E_t > W_{s_2} \tag{5.14}$$

The threshold condition for this case is determined by the following expressions:

$$\pi\frac{U_1}{U} = Arcth\, X_{s_2\uparrow E} - Arcth\, X_{s_1\uparrow E}$$
$$X_{s_2\uparrow E} = \frac{(e_{1z}+e_{3x})U_0 - W_{s_2} d(U_0/U_1)}{\pi K_{11}}$$
$$X_{s_1\uparrow E} = \frac{(e_{1z}+e_{3x})U_0 + W_{s_1} d(U_0/U_1)}{\pi K_{11}} \tag{5.15}$$

The solution of this transcendental equation is shown in Figure 5.6.

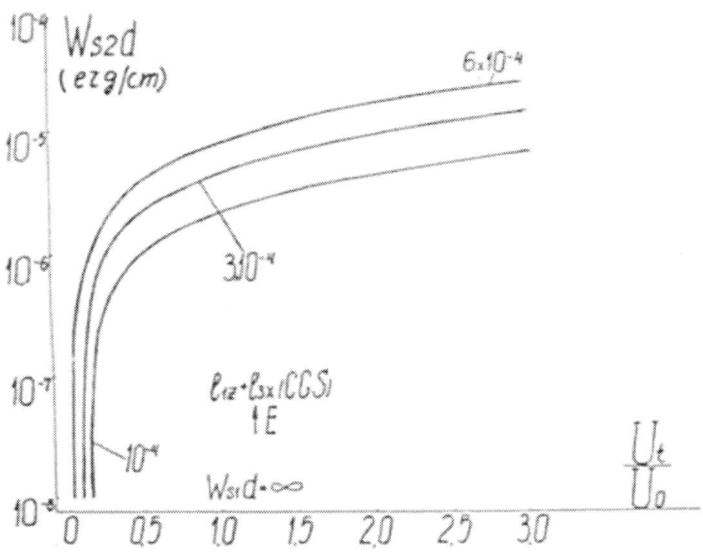

Figure 5.6. Polar theoretical curves at a voltage threshold for a homogeneous MBBA layer at various W_{s2}d and $W_{s1}d=\infty$.

The comparison of the curves shown in Figures 5.3-5.6 clearly demonstrates that at stabilizing action of the electric field, a larger electric field is needed for the appearance of the polar flexoelectric deformations. This study leads to the following important conclusion about the practical application of this effect for the case of nematics with small dielectric anisotropy: a) The value of the important parameter $W_s d \sim 3x10^{-7}$ erg/cm may be assumed for the lower limit of a rather weakly anchoring of the nematic liquid crystal; b) For the upper limit of a rather strongly fixed nematic, one may assume the value of $W_s d \sim 3x10^{-5}$ erg/cm; c) If the difference (in the pointed interval) between the two anisotropic surface energies is at least one order of magnitude, very clear distinctions are obtained in the thresholds for the polar flexoelectric effect.

5.3. On the "Gradient Flexoeffect" Once Again

In this part of the work we will focus our attention on new experimental results [105] obtained with non-homogeneous electric field, created flexoelectric deformations. We have used the MBBA layers, typically with a thickness of 20 μm with a small quantity of quinhydron (0.01 weight percent), added to the nematic phase (for details, see References [106-110]). This additive increased considerably the current from two times at 0.5 V to twenty times at 5 V when the liquid crystal was excitated by a DC voltage. We have observed the appearance of the gradient flexoelectric domains [57, 58] at 1-2 V when the glass plates of the initially planar nematic layers were treated by soap. This is the lowest threshold voltage, observed by us. Further, in the presence of quinhydron, the initial tilt of the nematic director increases up to $30 - 40^0$. When the quinhydron was removed, the initial orientation of the liquid crystal was nearly planar. The both differences can be simply explained by the reduction of the inverse electric field due to the double electric layers in the nematics. It is known that the quinhydron is a mixture from quinon $C_6H_4O_2$) and hydroquinon ($C_4H_4(OH)_2$). It is well-known that the hydroquinon oxidizes at the anode and two protons and two electrons are released. At the cathode, however, the quinon is transformed into quinhydron. The polarographic curves of the quinon and hydroquinon are completely reversible at 0.28 V. In addition, we know that water always persists in the liquid crystals [45] and it is possible to observe the following electrochemical reaction:

$$MBBA + H_2O \leftrightarrow pBA + pAde$$

Then the proton is attached to the nolecule of either MBBA or pBA and in this way are transformed into a protonated molecule, i.e., in a positive ion which flows to the cathode [110]. In this way, after a long working of the cell, the cathode is blocked by the positive ions

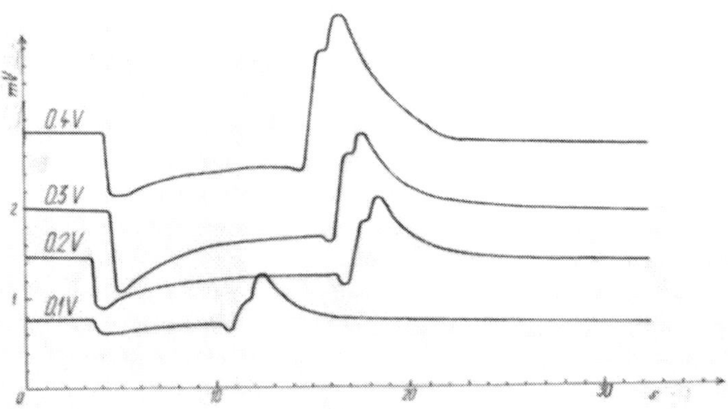

Figure 5.7. A typical gradient flexoelectric effect in MBBA cell with a thickness of 20 μm at an application of a very low DC voltage(0.01wt% quinhidron has been added). The maximum polarization of the Electrodes was at 0.4 V.

and the electric field inside the cell is strongly non-homogeneous. We show this process with observation of gradient flexoeffect at a very small voltage from 0.1 to 0.4 V (Figure 5.7).

It is seen that immediately after the disconnection of the voltage, the current increases considerably and we attribute this to flexoelectric gradient deformations.

5.4. On the Flexoeffect in Non-Homogeneous Electric Field due to Non-Planarity of the Cells and the Special Disposition of the Electrodes

Non-homogeneous electric field can also be created, as noted, by interdigital Electrodes [13, 72, 111-116]. Another source of non-homogeneity of the electric field can also be determined by the non-planar geometry of the cells [117-119]. In addition, conductor curvature creates surface charge density [120]. The use of Electrodes, disposed in another, more complex manner, also determines the non-homogeneity of the electric field [121-128]. There are also many papers which deal with the interaction between the non-homogeneous electric field and the nematic [129-132]. Finally, we should mention the methods for measurement of the electric field [133-135]. The bulk flexoelectric effect in a wedge-like nematic liquid crystal layer was considered in [136]. In this particular case, the electric field is between two glass plates. However, the electric field is additionally non-homogeneous due to the ions and because of that we have additionally applied low-frequency electric field. Our solution shows that the flexoelectric deformations dictated by bulk rotational torques are without thresholds. The orienting dielectric torques are without any threshold (negative value of the dielectric anisotropy). They act in the opposite direction which allows for the determination of some liquid crystal material constants. An investigation of this function shows that for MBBA the balance of the flexoelectric and dielectric torques takes place around 2.5 V and depends on the ratio $|\Delta\varepsilon|/K$, where K is the mean elastic constant.. The hydrodynamic effects blocking the flexoeffect begin at about 5 V.

The simpler case of a bulk flexoelectric effect in a wedge-like nematic, caused by a radial transversal electric field, was considered also by us [137]. In this case, the electric field is non-homogeneous along the coordinate r. Such electric field may be crudely created with the aid of two metallic (or conducting) sheets with different thickness (e.g., between 10 μm and 200 μm) or with two conducting wires. The distance between them will dictate the variation of the electric field along r. The distance between the two glass plates ensuring a strong anchoring of the nematic, should be smaller than 500 μm according to the results of Zwetkoff [138]. Otherwise, the thermal fluctuations of the nematic would prevail over the surface forces coming from the two glass plates. Then it is easy to measure exactly the total flexocoefficient even in a DC electric field for a large distance between the two conductors.

The theoretical calculations for the bulk flexoelectric deformations in wedge-like nematic cells considered here permit the determination of the sign and magnitude of the total flexoelectric coefficient ($e_{1z} + e_{3x}$) for nematics possessing either positive or negative dielectric anisotropy for the following two basic orientations of the liquid crystal.

a) Initial splay of the nematic liquid crystal with negative dielectric anisotropy.
b) Initial bend of the nematic liquid crystal with positive dielectric anisotropy.

The threshold conditions [139-140] lead to the following quadratic equation for the applied voltage U:

$$U^2 - \frac{8(e_{1z} + e_{3x})}{\Delta\varepsilon}\ln(r_a / r_b)U + \frac{\pi^3 K_{11} \ln^2(r_a / r_b)}{\Delta\varepsilon\varphi_0^2} = 0 \quad (5.16)$$

$$U^2 + \frac{8(e_{1z} + e_{3x})}{\Delta\varepsilon}\ln(r_a / r_b)U + \frac{\pi^3 K_{33} \ln^2(r_a / r_b)}{\Delta\varepsilon\varphi_0^2} = 0 \quad (5.17)$$

Equation (5.16) corresponds to the case (a), and equation (5.17) to the case b) and $\Delta\varepsilon$ can be either negative (case a) or positive (case b). The flexoelectric torques can be orienting or deforming and this depends on the sign of the total flexoelectric coefficient and the sign of the applied electric field. It is clear that at a given sign of $(e_{1z} + e_{3x})$ and $\Delta\varepsilon$ for the cases a) and b), there are two solutions of the quadratic equations (5.16) and (5.17) representing these two threshold voltages: U_1 and U_2. It is clear also that $(e_{1z} + e_{3x})$ can be obtained from the difference between these two threshold voltages:

$$U_1 - U_2 = \pm\frac{4\pi(e_{1z} + e_{3x})}{\Delta\varepsilon} ln(r_a / r_b) \quad (5.18)$$

The sign of the gradient of the electric field will determine the sign of $(e_{1z} + e_{3x})$ as well. This experiment must be performed only with deforming dielectric torques leading to the appearance of Fréedericksz threshold. The quadratic equations (5.16) and (5.17) show that the flexo-torques cannot overcome the stabilizing dielectric torques since they are much weaker. The knowledge of the sign of the electric field gradient can give the sign of the total flexoelectric coefficient as well. The estimation of MBBA shows that $\varphi_0 \sim 1\text{-}3^0$, $\ln(r_a/r_b) \sim 2\text{-}3$, $\Delta\varepsilon \approx 0.5$ are sufficient for the observation of a pronounced flexoelectric effect for $(U_1\text{-}U_2) \sim 10$ V and $U \sim 100\text{-}120$ V. The experiment should be carried out only with deforming dielectric torques for the investigated liquid crystal. Otherwise, the orienting dielectric torques would overcome the weak flexoelectric bulk torques. The other important feature is that the type of applied electric field should be close to the radial one. The hydrodynamic effect may be prevented at larger values of the wedge angle which should be $4\text{-}6^0$, thus decreasing sharply the threshold voltages. This method permits the determination of the value and the sign of the total flexocoefficient for a wide class of nematic liquid crystals.

5.5. Enhancement of the Flexoelectric Oscillations in Simultaneously Applied DC and Low-Frequency Electric Fields

During the experimental study of the "gradient flexoelectric effect", the author of this work has observed, for the first time, enhancement of the flexoelectric oscillations caused by a low-frequency electric field with a frequency up to 5 Hz when simultaneously with it also a DC voltage is applied [141]. It is clear, that at this set-up the observed effect is very complex since the exact distribution of the electric field is not known exactly inside the liquid crystal cell due to the presence of ions in the liquid crystal. This electro-optic effect can be observed without complications in the case when the two electric fields are applied transversely in the plane of the glass plates and the nematic layer is with homeotropic orientation of the director (an original proposal of Helfrich for measurement of the flexoelectric coefficient of bend e_{3x}). An original method for enhancement of the flexoelectric oscillations was proposed [142] on the basis of our preliminary observation and a simple theory was created. For the most simple case when only AC electric field is applied on the liquid crystal cell, the solution is of the kind:

$$\theta(z,t) = f_1(z) cos\, \omega t + f_2(z) sin\, \omega t \tag{5.19}$$

The intensity of the light transmitted through the liquid crystal cell is determined by the expression:

$$I(t) = I_0\, sin^2 \frac{2\pi}{\lambda} \frac{\delta_\theta}{2} \sim I_0 \left[\frac{\pi}{\lambda} \frac{n_0}{2} \left(1 - \frac{n_0^2}{n_e^2} \right) \int_{-d/2}^{+d/2} \theta^2(z,t)\, dz \right]^2 \tag{5.20}$$

In this expression, λ is the wave length and I_0 is the intensity of the incident light, n_e and n_0 are the optical indices of refraction of the liquid crystal with respect to the extraordinary and ordinary waves. The substitution of (5.19) in (5.20) makes clear that I(t) contains a constant term, second and fourth harmonics with respect to the applied AC electric field:

$$I(t) = I_0 + I_{2\omega} cos(2\omega t + \varphi) + I_{4\omega} cos(4\omega t + \psi) + ... \tag{5.21}$$

According to the calculations of Derzhanski [40], the amplitude of the second harmonic is the highest one and for its frequency dependence one obtains:

$$I_{2\omega} \sim E^4 \omega^{-3} \tag{5.22}$$

If we apply a DC voltage, we force the liquid crystal to oscillate around a deformed state. Let us the total electric field is expressed in the following manner:

$$E = E_0 + E_1 cos\, \omega t \tag{5.23}$$

Then the solution will be a linear combination of a solution which does not depend on the time and a solution depending on time:

$$\theta(z,t) = \theta_0(z) + \theta_1(z,t) \tag{5.24}$$

The displacement of the electro-optic system polarizer – liquid crystal layer – analyzer from the minimal transmission due to the initial deformation, leads to appearance of odd harmonics with weaker fall. We replaced the solution (5.24) for one concrete case and obtained in general form the following expression:

$$I(t) = I_0 + I_\omega \cos(\omega t + \varphi) + \ldots \qquad (5.25)$$

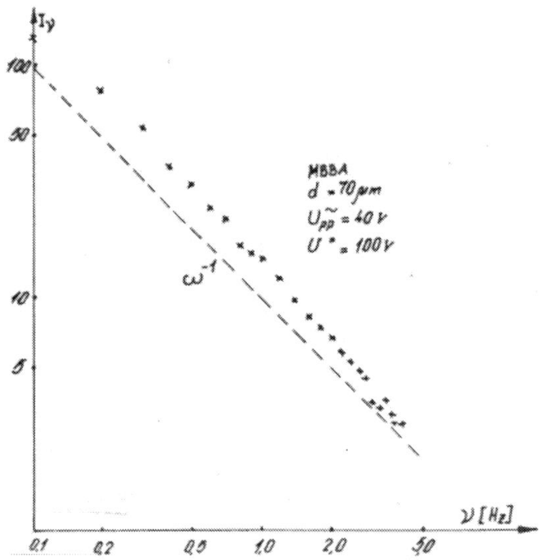

Figure 5.8. Flexoelectric oscillations with a varying frequency of the applied very low AC electric field in the presence of DC electric field for a MBBA nematic cell with a thickness of 70 μm.

According to the calculations of Derzhanski [40], the expression for I_ω is proportional to the third order of the applied d.c. electric field. His calculations show that the application of DC electric field leads to an abrupt increase of the amplitude of the basic harmonic, proportional to E_0^3 and that there is $1/\omega$ decrease of the intensity of the transmitted light as is shown in Figure 5.8.

Conclusion

1) The polar flexoelectric deformations are theoretically studied and the thresholds for various surface energies and flexoelectric coefficients are obtained for two cases: a deformation with a maximum inside the liquid crystal cells and a monotone deformation.
2) It is experimentally demonstrated a "gradient flexoeffect " in MBBA using an additive of quinhydron. The flexoelectric deformations were observed below 1 V.
3) It is theoretically proved that bulk flexoelectric effect arises in a wedge-shaped geometry and formulae for the determination of the magnitude and sign of total flexoelectric coefficient are obtained.
4) It is experimentally proved that the flexoelectric deformations can be identified when low-frequency (below 5 Hz) and DC voltages are simultaneously applied. The flexoeffect at such an experiment is with enhanced oscillation of the first harmonic.

REFERENCES

[1] Meyer, R. B. *Phys. Rev. Lett.* 1969, 22, 918-921.
[2] Meyer, R. B. In *Molecular Physics*; Balian, R.; Weil, G.; Eds.; Academic Press: New York, U.S. 1976; pp 271-344.
[3] De Gennes, P. G. *Physics of Liquid crystals*, Clarendon Press: Oxford, UK, 1974.
[4] Petrov, A. G. In *Physical Properties of Liquid crystals: Nematics*, emis Data Reviews Series N 25; Dunmur, D.; Fukuda, A.; Luckhurst, G.; Eds.; INSPEC, Publ. IEE, London, UK, 2001; pp 251-263.
[5] Fan, Ch. *Mol. Cryst. Liq. Cryst.* 1971, 13, 9-15.
[6] Derzhanski, A. I.; Petrov, A. G.; Mitov, M. D. *J. Phys.(Paris)* 1978, 39, 273-285.
[7] Pikin, S. A. *Structural Transformations in Liquid crystals*; Gordon & Breach: New York, U.S., 1991.
[8] Derzhanski, A. I.; Petrov, A. G. *Acta Phys. Polonica* 1979, 55, 747-767.
[9] Orsay Liquid crystal Group In *Liquid crystals & Ordered Fluids*; Griffin, A.C.; Johson, J. F.; Plenum Press: New York, London, U.S.,1970,; vol. 1, pp 195-200.
[10] Helfrich, W. *Mol. Cryst. Liq. Cryst.* 1973, 26, 1-5.
[11] Deuling, H. J. *Mol. Cryst. Liq.Cryst.* 1973, 26, 281-284.
[12] Deuling, H. J. *Mol. Cryst. Liq. Cryst.* 1973, 27, 81-93.
[13] Prost, J.; Pershan, P. S. *J. Appl. Phys.* 1976, 47, 2298-2312.
[14] Straley, J. P. *Phys. Rev. A* 1976, 14, 1835-1841.
[15] Marcerou, J.-P.; Prost, J. *Ann. Phys.(France)* 1978, 3, 269-272.
[16] Vasilevskaja, A. S.; Kaznacheev, A. B.; Sonin, A. S. *Fiz. Tv. Tela (In Russian)* 1982, 24, 3714-3715.
[17] Rudquist, P.; Lagerwall, S. T. *Liq. Cryst.(UK)* 1997, 23, 503-510.
[18] Derzhanski, A.; Petrov, A. G. *Phys. Lett. A* 1971, 34, 427-428.
[19] Helfrich, W. *Z. Naturforsch. (a)* 1971, 26, 833-835.
[20] Derzhanski, A.; Petrov, A. G.; Bivas, I. In *Advances in Liquid crystal Research and Applications*; Bata, L.;Ed.; Pergamon Press, Oxford-Akademiai Kiado, Budapest, Hungary,1980; pp505-513.
[21] Osipov, M. A. *Zh. Eksp. Teor. Fiz (USSR)* 1983, 85, 2011-2018.
[22] Osipov, M. A. *J. Phys. Lett.(Paris)* 1984, 45, L-823 – L-826.
[23] Singh, Y.; Singh, U. P. *Phys. Rev. A* 1989, 39, 4254-4262.
[24] Stelzer, J.; Herardi, R.; Zannoni, C. *Mol. Cryst. Liq. Cryst.* 2000, 352, 187-194.
[25] Billeter, J. L.; Pelcovits, R. *Liq. Cryst.(UK)* 2000, 27, 1151-1160.
[26] Ferrarini, A. *Phys. Rev. E* 2001, 64, 021710 (11 pp).
[27] Zakharov, A. V.; Dong, R. Y. *Eur. Phys. J. E* 2001, 6, 3-6.
[28] Madhusudana, N. V.; Chandrasekhar, S. *Pramana* 1975, suppl. N 1, 57-68.
[29] Maurel, P.; Price, A. H. *J. Chem. Soc. Far. Tr. II* 1975, 69, 1486-1490.
[30] Dunmur, D. A.; Miller, W. H. *Mol. Cryst. Liq. Cryst.* 1980, 60, 281-292.
[31] Megnassan, E.; Proutiere, A. *Mol. Cryst. Liq. Cryst.* 1984, 108, 245-354.
[32] Mason, W. P. *Piezoelectric crystals*; D. van Nostrand Company, Inc., Princeton, N. Y., U.S., 1950.
[33] Dubois-Violette, E.; Parodi, O. *J. Phys. Colloq.(Paris)* 1969, 30, C4-57 – C4-64.
[34] Lavrentovich, O. D. *Pis'ma Zh.T.Fiz.(USSR)* 1988, 14, N2, 166-171.

[35] Kurik, M. V.; Lavrentovich, O. D. *Izv. Ak. Nauk, Ser. Fiz.(USSR)* 1989, 53, 1880-1904.
[36] Derfel, G.; Felczak, M. *Proc. SPIE* 2002, 4759, 286-292.
[37] Hinov, H. P. *Cryst. Res. Technol.* 2009, 44, 327-330.
[38] Rapini, A. In *Progress in Solid State Chemistry*; McCaldin, J. O.; Somorjai, G.; Eds.; Plenum Press, N. Y., U.S., 1978, vol.8; pp337-401.
[39] Derzhanski, A. I.; Petrov, A. G.; Khinov, Chr. P.; Markovski, B. L. *Bulg. J. Phys.* 1974, 1, 165-174.
[40] Derzhanski, A. I. *Dr. Sci. Thesis* 1980, Sofia, Bulgaria.
[41] Onsager, L. *Science* 1969, 166, N 3911, 1359-1364.
[42] Heppke, G.; Schneider, F. *Chem. Phys. Lett.* 1972, 13, 548-551.
[43] Sprokel, G. J. *Mol. Cryst. Liq. Cryst.* 1973, 22, 249-260.
[44] Hakemi, H.; Labes, M. M. *J. Chem. Phys.* 1974, 61, 4020-4025.
[45] Komarov, V. S.; Rusalovich, A. I. *Izv. Ak. Nauk., Ser. Chim. USSR* 1978, N5, 9-14.
[46] Sussman, A. *J. Appl. Phys.* 1978, 49, 1131-1138.
[47] Koezuka, H.; Kanegae, H.; Ono, H.; Shibayama, K. *J. Appl. Phys.* 1982, 53, 496-501.
[48] Gusak, N. A.; Kamach, Yu. E.; Ovchinnikov, V. M.; Sotskii, B. A. *Dokl. Ak. Nauk BSSR* 1983, XXVII, 904-906.
[49] Thurston, R. N. *J. Appl. Phys.* 1984, 55, 4154-4161.
[50] Harris, L. B.; Fiasson, J. *J. Phys. E: Sci. Instrum.* 1984, 17, 788-792.
[51] Mada, H. *Jpn. J. Appl. Phys.* 1990, 29, L123-L125.
[52] Imai, M.; Naito, H.; Okuda, M.; Sugimura, A. *Mol. Cryst. Liq. Cryst.* 1995, 259, 37-46.
[53] Colpaert, C.; Maximus, B.; De Meyere, A. *Liq. Cryst.(UK)* 1996, 21, 133-142.
[54] Chen, P.-L.; Chen, Sh.-H. *Jpn. J. Appl. Phys.* 2000, 39, L297-L299.
[55] Sawada, A.; Sato, H.; Manabe, A.; Naemura, Sh. *Jpn. J. Appl. Phys.* 2000, 39, 3496-3501.
[56] Korniychuk, P. P.; Gabovich, A. M.; Singer, K.; Voitenko, A. I.; Reznikov, Yu. A. *Liq. Cryst.(UK)* 2010, 37, 1171-1181.
[57] Hinov, H. P.; Vistin', L. K. *J. Phys.(Paris)* 1979, 40, 269-292.
[58] Hinov, H. P.; Bivas, I.; Mitov, M. D.; Shoumarov, K.; Marinov, Y. *Liq.* Cryst.*(UK)* 2003, 30, 1293-1317.
[59] Lu, S.; Jones, D. *Appl. Phys. Lett.* 1970, 16, 484-486.
[60] Assouline, G.; Hareng, M.; Leiba, E. *IEEE Trans. El. Dev.* 1971, ED-18, 959-964.
[61] Meyerhofer, D.; Sussman, A. *Appl. Phys. Lett.* 1972, 20, 337-339.
[62] Goodisman, J. *J. Chim. Phys.* 1975, 72, 143-147.
[63] Yanagisawa, T.; Matsumoto, H.; Yahagi, K. *Jpn. J. Appl. Phys.* 1977, 16, 45-48.
[64] Yamashita, M.; Amemiya, Y. *Jpn. J. Appl. Phys,* 1978, 17, 1513-1517.
[65] Derfel, G.; Lipinski, A. *Mol. Cryst. Liq. Cryst.* 1979, 55, 89-100.
[66] Cognard, J. J. *J. Electrochem. Chem.* 1984, 160, 305-319,
[67] Derfel, G.; Felczak, M. *Liq. Cryst.(UK)* 2002, 29, 889-897.
[68] Felczak, M.; Derfel, G. *Liq. Cryst.(UK)* 2003, 30, 739-746.
[69] Hinov, H. P. *Bulg. J. Phys.* 2004, 31, 55-67.
[70] Brown, C. V.; Mottram, N. J. *Phys. Rev. E* 2003, 68, 031702 (5 pp).
[71] Prost, J.; Marcerou, J. P. *J. Phys.(Paris)* 1977, 38, 315-323.
[72] Marcerou, J. P.; Prost, J. *Mol. Cryst. Liq. Cryst.* 1980, 58, 259-284.
[73] Dozov, I.; Barbero, G.; Palierne, J. F.; Durand, G. *Europhys. Lett.* 1986, 1, 563-569.
[74] Parfenov, A. V.; Chigrinov, V. G. *Liq. Cryst.(UK)*1990, 7, 131-141.

[75] Chigrinov, V. *Mol. Cryst. Liq. Cryst.* 1990, 179, 71-91.
[76] Chigrinov, V.; Nevskaya, G. E. *Mol. Cryst. Liq. Cryst.* 1991, 209, 9-18.
[77] Smith, A. A. T.; Brown, C. V.; Mottram, N. J. *Phys. Rev. E* 2007, 75, 041704 (8pp).
[78] Petrov, A. G.; Derzhanski, A. *Mol. Cryst. Liq. Cryst. Lett.* 1977, 41, 41-46.
[79] Chuvyrov, A. N.; Lachinov, A. N. *Zh. Eksp. Teor. Fiz.(USSR)* 1978, 74, 1431-1444.
[80] Dowell, F. *Phys. Rev. A* 1985, 31, 3214-3219.
[81] Lavrentovich, O. D.; Pergamenshchik, V. M.; Sergan, V. V.; Sorokin, V. M. *Zh. Eksp. Theor. Fiz. (Russia) (in Russian)* 1991, 99, 777-802.
[82] Blinov, L. M.; Barnik, M. I.; Ozaki, M.; Shtykov, N. M.; Yoshino, K. *Phys. Rev. E* 2000, 62, 8091-8099.
[83] Blinov, L. M.; Barnik, M. I.; Ohoka, H.; Ozaki, M.; Yoshino, K. *Phys. Rev. E* 2001, 64, 031707 (7 pp).
[84] Mazzulla, A.; Ciuchi, F.; Sambles, J. R. *Phys. Rev. E* 2001, 64, 021708 (6 pp).
[85] Jungerman, J. A. *Rev. Sci. Instrum.* 1984, 55, 1479-1482.
[86] Helfrich, W. *Appl. Phys. Lett.* 1974, 24, 451-452.
[87] Derzhanski, A. I.; Hinov, H. P. *J. Phys.(Paris)* 1977, 38, 1013-1023.
[88] Madhusudana, N. V.; Durand, G. *J. Phys. Lett.(Paris)* 1985, 46, L-195 – L-200.
[89] Monkade, M.; Martinot-Lagarde, Ph.; Durand, G. *Europhys. Lett.* 1986, 2, 299-305.
[90] Barbero, G.; Durand, G. *Phys. Rev. A* 1987, 35, 1294-1300.
[91] Lee, S.-D.; Patel, J. S. *Phys. Rev. Lett.* 1990, 65, 56-59.
[92] Lee, S.-D.; Patel, J. S. *Mod. Phys. Lett. B* 1990. 4, 1071-1076.
[93] Lee, S.-D.; Rhee, B. K.; Jeon, Y. J. *J. Appl. Phys.* 1993, 73, 480-482.
[94] Kini, U. D. *Liq. Cryst.(UK)*1993, 13, 735-755.
[95] Lee, J.; Suh, S.-W.; Lee, K.; Lee, S.-D. *Phys. Rev. E* 1994, 49, 923-926.
[96] Klouda, R.; Nazarenko, V. G.; Kosa, T.; Hsia, K.-M.; Lavrentovich, O. D.; Palffy-Muhoray, P. *Mol. Cryst. Liq. Cryst.* 1997, 304, 133-138.
[97] Kim, Y. J.; Suh, S.-W.; Lee, S.-D. *Mol. Cryst. Liq. Cryst.* 1997, 302, 169-174.
[98] Wu, Ch.-X.; Iwamoto, M. *Phys. Rev. E* 1998, 57, 5740-5743.
[99] Nazarenko, V, G.; Klouda, R.; Lavrentovich, O. D. *Phys. Rev. E* 1998, 57, R36-R38.
[100] Ponti, S.; Ziherl, P.; Ferrero, C.; Zumer, S. *Liq. Cryst.(UK)* 1999, 26, 1171-1177.
[101] Forget, S.; Dozov, I.; Martinot-Lagarde, Ph. *Mol. Cryst. Liq. Cryst.* 1999, 329, 605-615.
[102] Reznikov, Y.; Buchnev, O.; Tereshchenko, O.; Reshetnyak, V.; Glushchenko, A.; West, J. *Appl. Phys. Lett.* 2003, 82, 1917-1919.
[103] Liu, J.; Zhang, S.; Yang, Y.; An, H.; Zhang, Z.; Yang, G. *Liq. Cryst.(UK)* 2007, 34, 1425-1431.
[104] Barratt, P. J.; Jenkins, J. T. *J. Phys. A: Math. Nucl. Gen.* 1973, 6, 756-769.
[105] Hinov, H. P., unpublished results.
[106] Glasstone, S. *Introduction to Electrochemistry,* N.Y., Translated in Russian by Inostranoi Literature, Moscow, 1951; pp 380-393.
[107] Frumkin, A. N.; Bagotskii, V. S.; Iofa, Z. A.; Kabanov, B. N. *Kinetics of Electrode Processes*; Izd. Moscow University, 1952, chapter II.
[108] Ohnishi, Y.; Ozutumi, M. *Appl. Phys. Lett.* 1974, 24, 213-216.
[109] Kohlmüller, H.; Siemsen, G. *Siemens Forsch. U. Entwickl-Ber.* 1982, 11, 229-234.
[110] Brière, G.; Herino, R.; Mondon, F. *Mol. Cryst. Liq. Cryst.* 1972, 19, 157-177.
[111] Blümich, B. P.; Germer, R. *J. Phys. E: Sci. Instrum.* 1978, 11, 114-115.

[112] Lindquist, R. G.; Kulick, J. H.; Nordin, G. P.; Jarem, J. M.; Kowel, S. T.; Friends, M.; Leslie, T. M. *Optics Lett.* 1994, 19, 670-672.

[113] Schnelle, Th.; Müller, T.; Voigt, A.; Reimer, K.; Wagner, B.; Fuhr, G. *Langmuir* 1996, 12, 801- 809.

[114] Lee, S. H.; Lee. S. L.; Kim, H. Y. *Appl. Phys. Lett.* 1998, 73, 2881-2883.

[115] Chen, C.-J. *J. Appl. Phys.* 2000, 87, 5-9.

[116] Ueno, K.; Hayashida, M.; Ye, J.-Y.; Misawa, H. *Electrochem. Commun.* 2005, 7, 161-165.

[117] Klopfenstein, R. W.; Wehner, R. K. *RCA Review* 1973, 34, 630-654.

[118] Chute, F. S.; Vermeulen, F. E. *AJP* 1974, 42. 1075-1077.

[119] Boridy, E. *IEEE Tr. Electromagnetic Comput.* 1987, EMC-29, 132-140.

[120] McAllister, I. W. *J. Phys. D: Appl. Phys.* 1990, 23, 359-362.

[121] Durand, E. *Electrostatique I Les distributions*, Masson et Cie; 1964, pp 170-173, pp 219-221.

[122] Durand, E. *Electrostatique II Problems genèraux conducteurs,* Masson et Cie; 1966, pp 239-245.

[123] Sakudo, N.; Hayashi, T. *Rev. Sci. Instrum.* 1975, 46, 1060-1062.

[124] Blankson, I. M. *J. Electrostatics* 1984, 16, 33-45.

[125] Elmoursi, A. A.; Castle, G. S. P. *J. Electrostatics* 1987, 19, 221-234.

[126] Hechtman, Ch. D.; Levine, Z. H. *J. Appl. Phys.* 1987, 62, 2116-2121.

[127] Jacquelin, J. *Electrochimica Acta* 1993, 38, 597-606.

[128] Homentcovschi, D. *J. Electrostatics* 1991, 26, 187-200.

[129] Chigrinov, V. G.; Kompanetz, I. N.; Vasilev, A. A. *Mol. Cryst. Liq. Cryst.* 1979, 55, 193-208.

[130] Yoon, T.-Y.; Park, J.-H.; Sim, J.-S.; Lee, S.-D. *Appl. Phys. Lett.* 2002, 81, 2361-2363.

[131] Xiang, C. Y.; Sun, X. W. *J. Appl. Phys.* 2004, 96, 3520-3523.

[132] Davidson, A. J.; Elston, S. J.; Raynes, E. P. *J. Phys. D: Appl. Phys.* 2005, 38, 376-384.

[133] Jachym, A.; Jachym, B.; Tomaszewicz, W.; Zielinski, R. *Acta Phys. Polonica A*, 39, 441-451.

[134] Kulyupin, Yu. A.; Nepijko, S. A.; Sedov, N. N., Shamonya, V. G. *Optik* 1978/1979, 52, 101-109.

[135] Sueda, H.; Kao, K. C. *Appl. Optics* 1979, 18, 3990-3995.

[136] Derzhanski, A. I.; Hinov, H. P.; Mitov, M. D. *Acta Phys. Polonica* 1979, 55, 567-573.

[137] Hinov, H. P.; Derzhanski, A. I. In *Advances in Liquid crystal Research and Applications*; Bata, L.; Ed.; Pergamon Press, Oxford-Akademiai Kiado, Budapest, Hungary 1980; pp 523-526.

[138] Tsvetkov, V. *Ph.D.Thesis,* Institut of Radioelectronics, Moscow, 1974.

[139] Frank, F. C. *Discuss. Faraday Soc.* 1958, 25, 19-28.

[140] Leslie, F. M. *J. Phys. D: Appl. Phys.* 1970, 3, 889-897.

[141] Hinov, H. P., unpublished results.

[142] Petrov, A. G.; Derzhanski, A, I,; Avramova, K. S.; Hinov, H. P. *A Method for Obtaining and Enhancement of Flexoelectric Oscillations of the Light which Passes through the Nematic Liquid crystals*, Certificat for authorship N 26915, Reg. N 40481/78 1978, 1-7.

Chapter 6

DC VOLTAGE-CONTROLLED PULSATIONS OF A HOMEOTROPIC NEMATIC LAYER WITH A FREE-SLIPPING LIQUID CRYSTAL-POLAR LIQUID BOUNDARY

There are papers, in which the authors consider the possibilities for nonlinear oscillations of either polar liquids [1] or liquid crystals [2, 3]. We have studied for the first time the system semidrop of liquid crystal, embedded into a polar liquid or vice-versa, semi-drop of a polar liquid, embedded into a liquid crystal with a DC voltage applied across this system [4] (Figure 6.1).

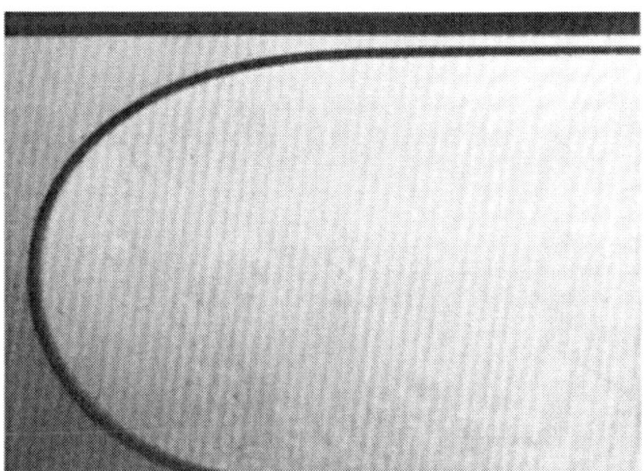

Figure 6.1. A schematic view of the liquid crystal semi-drop embedded into a polar liquid or vice-versa, semi-drop of a polar liquid embedded into the liquid crystal.

The liquid crystal cell was described in chapter 1. It consists of two glass plates with a large area (several cm^2) with a thickness between 10 and 100 μm. The glass plates were fixed by two lateral massive metal holders which ensure the electrical contact as well. The mechanical stability was achieved by means of four screws. The liquid crystal was introduced in the empty cell in its isotropic phase. Due to the capillary forces, the liquid crystal filled nearly the whole gap which was formed between the glass plates, separated by the polymer spacers with a different thickness. The second step in the preparation of the liquid crystal samples was the addition of a polar liquid such as water [5, 6] or glycerin [7], or a non-polar liquid such as transformer oil. The polar or non-polar liquid was added laterally from one side of the liquid crystal cells by a droppler. We suggested that part of the polar or non-polar liquid replaces the boundary liquid crystal layer where it wets more efficiently the glass plates covered by the semi-transparent conductive layer. The schematic representation of the experimental set-up is shown in Figure 6.2.

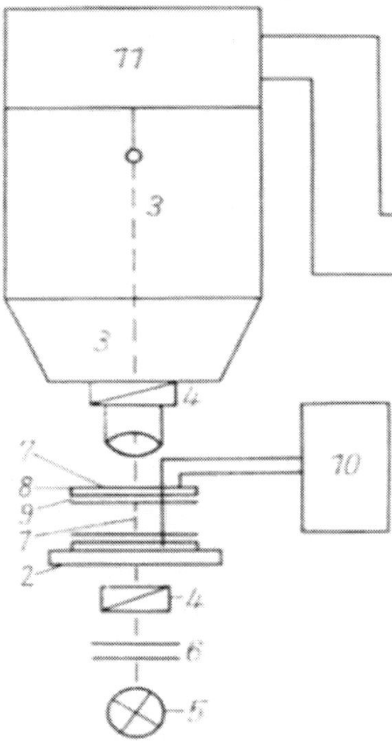

Figure 6.2. A schematic representation of the experimental set-up with details, designated as follows: 1. liquid crystal; 2. microscope stage; 3. polarizing microscope; 4. nicols; 5. light source; 6. monochromatic filter; 7. glass plates; 8. semitransparent conductive layer; 9. polar liquid; 10. voltage source; 11. photomultiplier; 12. recorder.

It is seen that in front of the liquid crystal there is a polar or non-polar liquid (compare with schematic picture of Shonova and Zheliaskova, presented for the same problem [8, 9-16]. Evidently, the treatment of the glass plates ensures their hydrophilic or hydrophobic character and solves the boundary problems [17-21]. The transformer oil was not able to replace the liquid crystal boundary layer and we have observed a clear phase boundary between the liquid crystal and the transformer oil added. We did not perform experimental measurements of the thickness of the polar liquid sheets since they require special technical means (for instance, a total internal reflection set-up). Nevertheless, we can approximately accept that this thickness is below 1 μm. The liquid crystal studied was pentyl-cyano-biphenyl (5CB) with a small amount of Lecithin (1 weight percent) added to ensure the homeotropic alignment of the nematic director. Additionally various nematics such as MBBA, a mixture of MBBA and 5CB, etc. were studied as well. The conductivity of the nematic 5CB was measured to be $5x10^{-10}$ $\Omega^{-1}cm^{-1}$. Another very important measurement was the conductivity of the pure glycerin which turned to be $1.2x10^{-9}$ $\Omega^{-1}cm^{-1}$. We have performed the experiment with pure glycerin whose chemical formula is $CHOH(CH_2OH)_2$, by the experimental measurements shown by the observed pulsations of the light transmitted through the liquid crystal cell [22, 23].

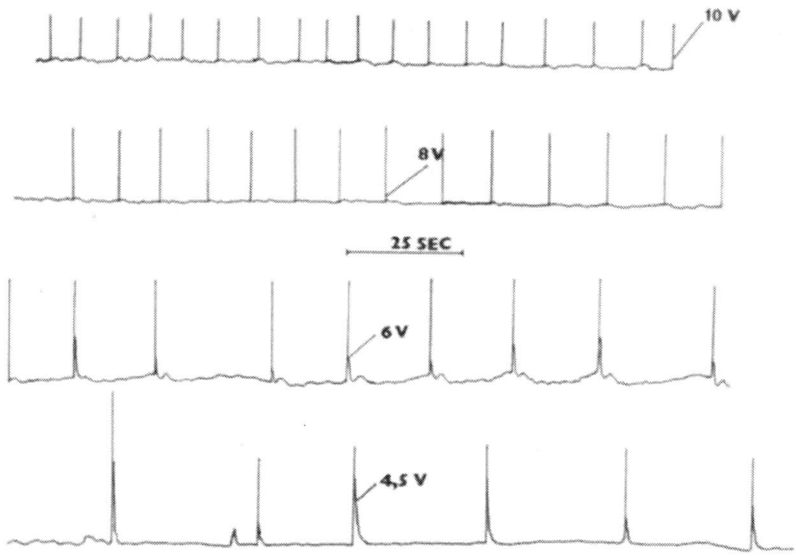

Figure 6.3. DC voltage-Controlled Pulsations in a 5CBliquid crystal-glycerin layer with a thickness of 20 μm subjected to the action of a DC voltage of 4.5, 6, 8 and 10 V.

In effect, this is a peculiar focusing of the light. One realizes that the pulsations start with a voltage threshold of 4.5 with a 25 s separation between every two flashes. The pulsations of the liquid crystal layer were quasi-periodic with infrasound frequency and were very well defined. The dielectric constants, parallel and perpendicular to the nematic director have been measured, for instance in [24]. [25]. Important material parameters are also the optical indices of the nematic 5CB [26, 27]. The visco-elastic properties of 5CB are also of importance for the estimation of the observed pulsations of the light [28] (some important material parameters for 5CB can be found also in Reference [29]). Two types of glycerin were used: a very pure glycerin with measured conductivity of σ in the range of $1.2x10^{-9}$ $\Omega^{-1}cm^{-1}$ a value which is near to that reported by Kukoina $1.6x10^{-9}$ $\Omega^{-1}cm^{-1}$at 25^0 C [7] and a more conductive glycerin(see below). Let us point out that we have performed additional experiments in order to understand the physics of the pulsations of the nematic layer. First, we used glass plates with a special treatment of Lecithin (below 0.1 weight percent Lecithin in alcohol). It is well-known that the Lecithin has a hydrophilic character while the glycerin is hydrophobic. Indeed, the glycerin posed on the so-prepared glass plate does not wet the glass plate well and many glycerin droplets were formed as is shown in Figure 6.4.

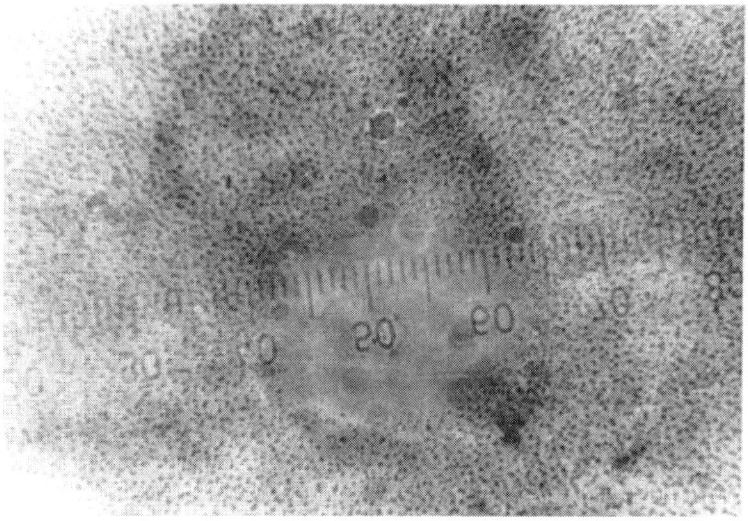

Figure 6.4. Many glycerin small droplets on one hydrophil surface of the glass plate.

Then we have prepared a liquid crystal cell, posing first the liquid crystal 5CB followed by the glycerin laterally. The conductivity of the glycerin was measured to be $1x10^{-8}\ \Omega^{-1}cm^{-1}$. Under a DC voltage excitation, this liquid crystal cell shows also pulsations of the light transmitted through the liquid crystal cell. However, when the voltage grows up to 5 V, we have observed a strong electrolysis which destroys the glycerin with formation of many holes filled with hydrogen near the cathode and the current was measured to be above 80 μA (see Figure 6.5).

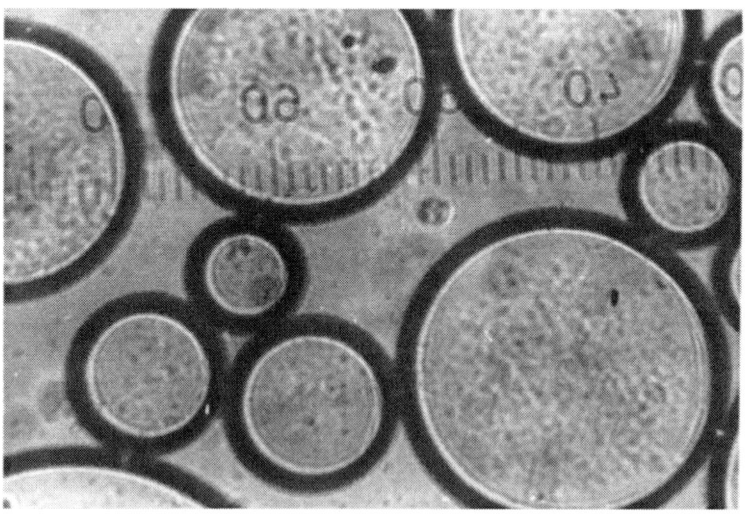

Figure 6.5. Liquid crystal-glycerin layer under a DC voltage of 5V. It is seen many holes with different sizes in the glycerin filled with hydrogen.

At this voltage (5 V) we have observed the formation of circular-like electro-hydrodynamic domains of Felici shown in Figure 6.6 [30, 31].

Let us point out that the electrolysis is strongly temperature-dependent, For instance, at 18-19^{0}C it is rather weak [32] while at a room temperature of 25^{0}C, it is quite intensive. Another polar liquid, as already noted, which was used in our experiment, was water [33-39]. The electrolysis and the breakdown of water has been considered in References [40] and [41]. Additional tested nematic, as noted, was MBBA [42, 43].

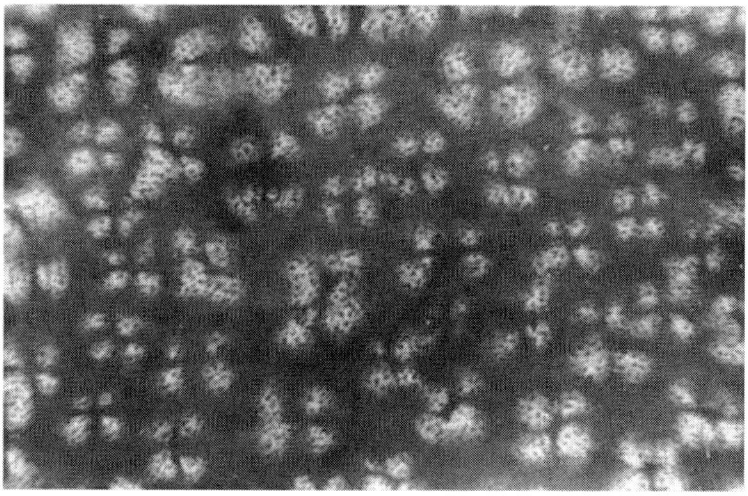

Figure 6.6. Liquid crystal-glycerin layer under a DC voltage of 5V. One sees many cross-like electro-hydrodynamic injection domains of Felisi in the nematic phase of 5CB.

The observed phenomenon is very complex and there are several ways for clarification. There are two mechanisms which could explain the observed effect. The first one is connected with the formation of double electric layers at the boundaries, first for the case of water [42-49] and second, for the case of the liquid crystal [50-55]. In addition, the hydrophilic or hydrophobic character of the glass plates used in the construction of the samples due to their treatment with surface-active substances is also of a crucial importance for the observation of this effect [49]. Our experiments clearly show that the glycerin ($\sigma \sim 1{,}56 x 10^{-8}\ \Omega^{-1}cm^{-1}$) has a hydrophobic nature while the non-polar liquid (Triton X 114, $\sigma \sim 1{,}56 x 10^{-8}\ \Omega^{-1}cm^{-1}$) and the nematic MBBA showed hydrophilic character. In addition, the conductivity of highly purified water has been measured to be $10^{-7} - 10^{-8}\ \Omega^{-1}cm^{-1}$ [5, 34]. The conductivity of the usual non-purified water which was used in our experiment evidently is one or two orders of magnitude higher. As noted, the conductivity of the glycerin was measured to be lower (one or two orders of magnitude) in comparison to that of water. The dielectric constant of water is in the range of 80 [5, 30] and that of glycerin in the range of 42 [7]. Analogically, the conductivity of the nematic 5CB was measured to be in the range of that of the glycerin while the conductivity of MBBA is slightly higher ($\sigma \sim 7.8 x 10^{-9}\ \Omega^{-1}cm^{-1}$).

The liquid crystal cell was placed on the microscopic stage of the polarizing microscope between crossed nicols in the transmitted light of the light source which passes through the monochromatic filter and the liquid crystal. The glass plates of the liquid crystal cells were covered with a semitransparent conductive layer. We suggested initially that the liquid crystal layer is separated from the walls by two very thin sheets of the polar liquid. The so-prepared cells were excited by DC voltage. The electro-optical response of the liquid crystal was amplified by a photomultiplier and recorded by a X-Y recorder. As a probable reason for appearance of the pulsations we can suggest first the electro-optical behavior of the two very thin polar liquid-liquid crystal layers [50-63]. The second idea is connected with the role of the electrolysis which generates a pressure pushing the nematic semi-drop [54-70]. Let us stress that the pulsations of the light in all cases begin a 3-4.5V DC voltage (Figure 6.7).

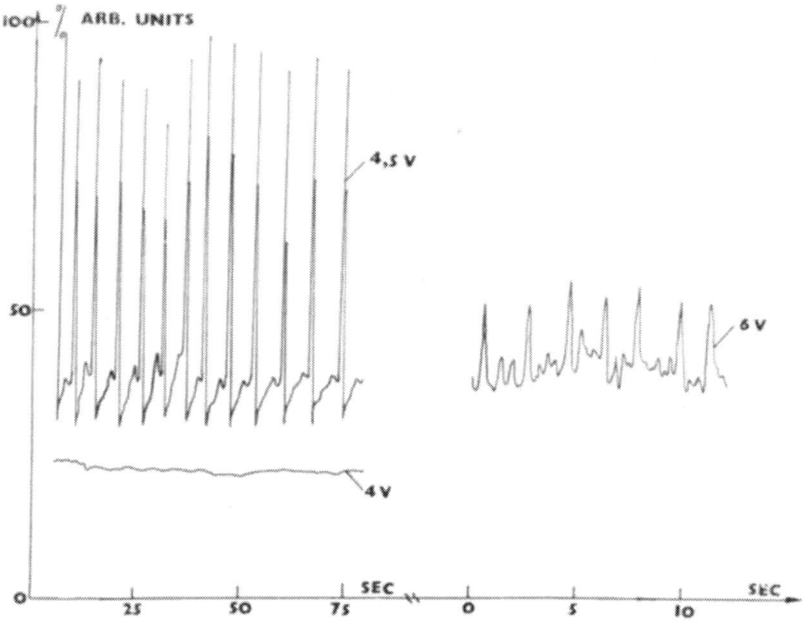

Figure 6.7. DC voltage-Controlled Pulsations in a 5CB nematic liquid crystal – water layer with a thickness of 20 μm. The pulsations appear at 4.5V. The electro-hydrodynamic domains appear at 6V.

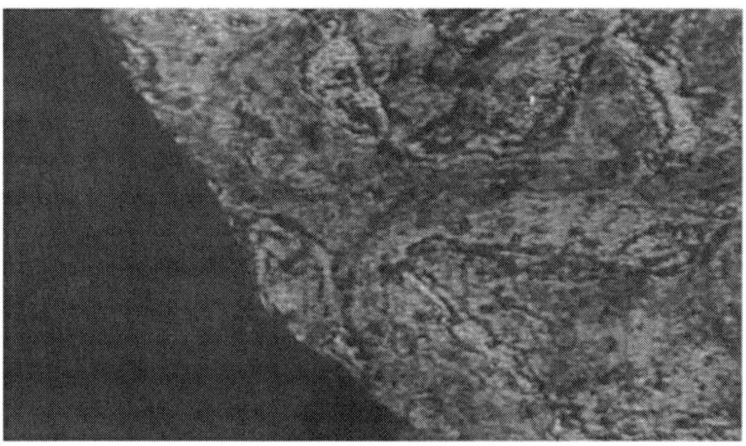

Figure 6.8. Many holes formed in the MBBA nematic after the pulsations of the layer.

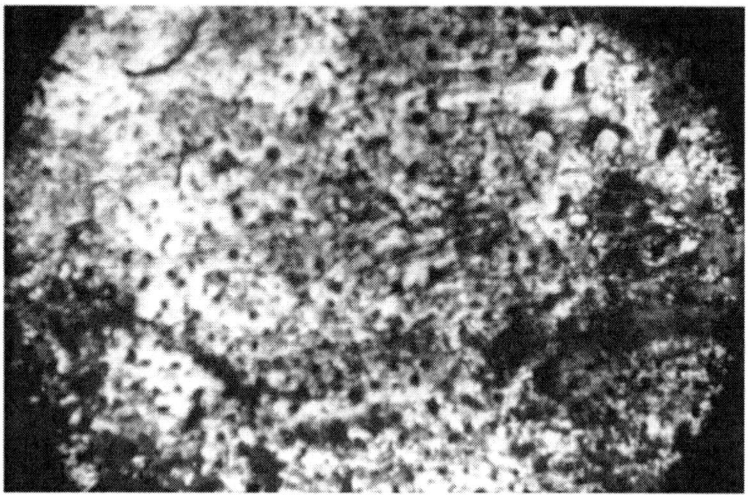

Figure 6.9. Many holes formed in the 5CB nematic after the pulsations of the layer.

They saturate (with increasing frequency of pulsations) at 6-10 V depending on the conductivity of the polar liquid and on the conductivity of the liquid crystal. We have not observed a degradation of the liquid crystal [71]. However, we have observed many small holes in the liquid crystal layers both for the cases of MBBA or 5CB (Figure 6.8 and Figure 6.9), similar to those observed by Yokoyama et al., [72] as well as degradation in the conductive layer of the semi-transparent Electrodes (this effect can be avoided by use of electrochemical substances as quinhydron or ferrocene with very low oxcidation potentials).

In order to understand the causes leading to the appearance of the pulsations we cut the electrically semi-transparent layer in order to prevent the electric connection to the liquid crystal. Then the DC voltage was applied only to the glycerin layer. Surprisingly, the pulsations of the light above the threshold voltage again reappear. They stopped when we replaced the polar liquid with non-polar liquid such as Triton X 114. However, with the system glycerin-Triton X 114-liquid crystal they again reappear, only with a smaller frequency and smaller amplitude. It is clear that the non-polar liquid Triton X 114 surveys as a buffer for the pulsations. These experiments clearly showed that our initial suggestion that the glycerin replaces the liquid crystal at the boundaries is wrong. Evidently the glycerin cannot replace the liquid crystal at the boundaries and in effect, we have a semi-drop of polar or non-polar drop embedded into the liquid crystal (see Figure 6.1). It is well- known that the glycerin has a hydrophobic character and 5CB has a hydrophilic character. As noted by Lavrentovich and Nastishin [73], the nematic 5CB is oriented planarly on the glycerin. Consequently, in the boundary regions we have hybrid nematic layer with homeotropic orientation at the boundary and a planar orientation at the glycerin. This suggestion confirms that the boundary of the glass plates-nematic-polar liquid is very important for the creation of pulsations of light. Further, the viscosity behavior of the nematic, describing by Rey [74-77] and the surface viscosity [78-102] are also of great importance for the appearance of pulsations.

Let us point out that at the pulsations, the field of view under the polarizing microscope, suddenly became bright for a fraction of a second. This pulsation was repeated many times with a period which for water was about 3 to 4 seconds while for glycerin it increases up to

30 seconds. The electro-optical pulsations can be created only in relatively thin liquid crystal layers with a thickness below 100 μm. We did not observe such pulsations in relatively thick liquid crystal layers with a thickness of 100 μm and above. In thick cells, we observed only slight pulsations due to the electro-hydrodynamic domains being formed. The pulsations were also studied in a conoscopic regime of the passing light. Since the changes in the conoscopic images were very fast, we were not able to discuss their exact kind without utilization of a special set-up. Let us point out that the whole liquid crystal pulsates on a very large area in the range of cm^2. Further, no domains were observed up to 6 V for the case of water and 10 V for the case of glycerin. The appearance of various kinds of domains can only disturb the electro-optical pulsations without a possibility to prevent them. The mean current through the liquid crystal layer and the polar liquid layer was relatively large (about 1.5 mA and during the pulsation it increases by 0.05 mA). It was difficult to draw a parallel between the pulsations of the liquid crystal director and those of the current since the current pulsations are integral whereas the pulsations of the liquid crystal director were investigated and recorded in the field of view of the polarizing microscope. Furthermore, in many cases we have observed several large bubbles of the polar liquid which under a DC voltage excitation disturbed and modified the current pulsations. Finally, the edges and the lateral boundaries were very often easily movable which also contributed to the current pulsations. No sparks have been observed during the experimental study of many liquid crystal-polar liquid samples.

The liquid crystal-polar liquid pulsations can be observed in homeotropic or planar nematics with either positive or negative dielectric anisotropy when various Surfactants are added in small (below 1 weight percent) or high (around 8 weight percent) amounts, with the formation of a variety of domains. The similarity in all these cases is, first, that such pulsations can be obtained above a threshold voltage in a nematic, and, second, that the period of these pulsations decreases with the increase of the voltage, third, the time of their reappearance depends on the nature of the polar liquid. It seems that the appearance of this instability leads to a discharge or to a destruction of the electric double layers forming at the two boundaries of the glass plates, This idea is supported by the following reasons: the threshold voltage for all the polar liquids added to the liquid crystal and for all thicknesses of the liquid crystals coincides with that for the creation of the double electric layers [54]. The thickness of the liquid crystal-polar liquid electric double layer depends strongly on the concentration of impurities [52] and typically is in the range of 50 to 500 nm [51, 52]. The voltage across the electric double layer is about 1 V [51, 52, 54]. On the other hand, the high electric field which around the threshold is between 1500 V/cm and 5000 V/cm depending on the thickness of the liquid crystal-polar layers permits the creation of ionic carriers in the polar liquid sheets by injection from the Electrodes. Here we should mention that the weak electrolysis is crucial for the creation of the pulsations, creating hydrogen in the cathode region. This assumption is confirmed by the considerable change in the semi-transparent conductive layers as well. Here we must mention that the mobility of the space charges in the polar liquids is much greater than the mobility of the space charges in the liquid crystal. For instance, the mean mobility of the ions in water is about $4x10^{-4}$ cm^2/Vs [34] while the mobility of the ions in the liquid crystal is around 10^{-5} to 10^{-6} cm^2/Vs [53]. It is clear that for the formation and saturation of the electric double layers some time is needed which for the case of liquid crystal-water is several seconds while for the case of liquid crystal-glycerin – several tens of seconds. It is more difficult to explain which physical mechanisms (except the electrolysis) are responsible for the appearance of the instability leading to the apparent discharge or to the destruction of the electric double layers. A dielectric breakdown of the material inside the electric double layers containing both the liquid crystal and the polar liquid, is less probable. For instance, the dielectric breakdown of a pure liquid crystal with a conductivity of $10^{-11}\ \Omega^{-1}cm^{-1}$ occurs at about 100 kV/cm [5, 71]. The value of the electric field which leads to a dielectric breakdown of the polar liquid evidently is of the same order or may be larger [41]. The discharge of the electric double layer probably is connected with electro-osmotic forces which take into account the charge density, the electrical potential, the current density, the pressure, and the vertical velocity of the fluid [1]. The physical picture is complicated by the existence of an interface between the liquid crystal and the polar liquid. Since the electrical double layers are near the Electrodes, the vertical movement of the interface can also lead to a discharge of part of the interfacial charges. At the present stage of the experimental evidence, we only know that the electro-optical pulsations of the liquid crystal-polar liquid layer are perpendicular to the glass plates of the cells. In other words, they are parallel to the direction of the electric field applied across the samples. This experimental finding was revealed by a Free-Slipping Liquid Crystal-Polar liquid Boundary which at tangential orientation of the liquid crystal molecules moves in the plane of the Electrodes. At the pulsations, however, this movement stops which clearly confirms that the direction of the forces is perpendicular to the plane of the Electrodes. The formation of pores or channels by unknown mechanisms in the polar liquid sheets and in the liquid crystal might facilitate the discharge of the electric double layers.

CONCLUSION

We have prepared for the first time a cell composed of a liquid crystal layer and a polar liquid penetrating each other. We have called, for convenience, the liquid crystal-polar liquid interface a free-slipping interface although in certain cases the liquid crystal molecules can be strongly attached to the polar liquid molecules. These new boundary conditions can lead to the observation of many novel electro-optical and other effects which can be usedin the liquid crystal devices. We made an attempt to explain qualitatively the physical causes eventually responsible for the creation of the observed new pulsations of the light passing through the samples. Nevertheless, additional experimental and theoretical work is needed for the clear explanation of this interesting and useful electro-optic effect.

REFERENCES

[1] Suzuki, M.; Sawada, Y. *J. Appl. Phys.* 1980, 51, 5667-5674.
[2] Cuculescu, I.; Bena, R.; Florea, C.; Savin, O.; Euache, A. *Phys. Rev. A* 1983, 27, 1744-1746.
[3] Goldburt, E. S.; Russell, P. St. *Opt. Lett.* 1986, 11, 51-52.
[4] Hinov, H. P. *Phys. Stat. Sol. (a)* 1987, 101, 185-192.
[5] Honda, T.; Sasada, T. *Jpn. J. Appl. Phys.* 1979, 18, 675-676.
[6] Liu, S. H. *Surf. Sci.* 1981, 105, 429-437.

[7] *Tables of Physical Quantities*; Kukoina, U. K.; Ed.; Gasatomizdat, Moscow, Russia, 1976.
[8] Shonova, N.; Zheliaskova, A. *Cryst. Res. Technol.* 1990, 25, K185-K189.
[9] De Gennes, P. G. *Rev. Mod. Phys.* 1985, 57, 827-863.
[10] Slukin, T. J.; Poniewierski, A. In *Fluid Interfacial Phenomena*; Croxton, C. A.; Ed.; John Wiley and Sons Ltd., 1986; pp 215-253.
[11] Cognard, J. *J. Chim Phys.1986*, 83, 541-547.
[12] Widom, B.; Clarke, A. S. *Physica A* 1990, 168, 149-159.
[13] De Gennes, P. G. In *Fundamentals of Adhesion*; Lee, L.-H.; Ed.; Plenum Publ. Corp.: New York, U.S. 1991; pp 173-179.
[14] Rey, A. D. *J. Chem. Phys.* 1999, 111, 7675-7684.
[15] Rey, A. D. *Liq. Cryst.(UK)* 2000, 27, 195-200.
[16] Poulard, C.; Cazabat, A. M. *Langmuir* 2005, 21, 6270-6276.
[17] Ericksen, J. L. *Trans. Soc. Rgeology* 1969, 13, 9-15.
[18] Fisher, J.; Fredrickson, A. G. *Mol. Cryst. Liq. Cryst.* 1969, 8, 267-284.
[19] Proust, J. E.; Ter-Minassian-Saraga, L.; Guyon, E. *Sol. St. Commun.* 1972, 11, 1227-1230.
[20] Williams, R. *Mol. Cryst. Liq. Cryst.* 1976, 35, 349-351.
[21] Derjaguin, B. V.; Popovskii, Yu. M.; Altoiz, B. A. *J. Coll. Int. Sci.* 1983, 96, 492-503.
[22] Kowel, S. T.; Cleverly, D. S.; Kornreich, Ph. G. *Appl. Optics* 1981, 23, 278-289.
[23] Nose, T.; Sato, S. *Liq. Cryst.(UK)* 1989, 5, 1425-1433.
[24] Ratna, B. R.; Shashidhar, R. *Mol. Cryst. Liq. Cryst.* 1977, 42, 113-125.
[25] Schell, K. T.; Porter, R. *Mol. Cryst. Liq. Cryst.* 1990, 188, 97-104.
[26] Wu, Sh.-T.; Wu, Ch.-Sh.; Warenghem, M.; Ismaili, M. *Opt. Eng.* 1993, 32, 1775-1780.
[27] Brugioni, S.; Faetti, S.; Meucci, R. *Optichesski Zhurnal (Russia)(In Russian)* 2006, 73, N 5, 15-17.
[28] Cui, M.; Kelly, J. R. *Mol. Cryst. Liq. Cryst.* 1999, 331, 49-57.
[29] Yokoyama, H.; Kobayashi, Sh.; Kamei, H. *Mol. Cryst. Liq. Cryst.* 1984, 107, 311-331.
[30] Hinov, H. P.; Vistin', L. K. *J. Phys.(Paris)* 1979, 40, 269-292.
[31] Hinov, H. P.; Bivas, I.; Mitov, M. D.; Shoumarov, K.; Marinov, Y. *Liq. Cryst.(UK)* 2003, 30, 1293-1317.
[32] Hinov, H. P., unpublished results.
[33] Stanley, H. E. *Pramana* 1999, 53, 53-83.
[34] Zahn, M.; Takada, T. *J. Appl. Phys.* 1983, 54, 4762-4775.
[35] Alekseev, O. L.; Ovcharenko, F. D. *Dokl. Akad. Nauk USSR* 1987, 292, 881-885.
[36] Labib, M. E.; Williams, R. *J. Coll. Int. Sci.* 1987, 115, 330-338.
[37] Pallas, N. R.; Pethica, B. A. *Colloids & Surfaces* 1989, 36, 369-372.
[38] Robinson, G. W.; Cho, Ch. H.; Gellene, G. I. *J. Phys. Chem. B* 2000, 104, 7179-7182.
[39] Zhu, Y.; Granick, S. *Phys. Rev. Lett.* 2001, 87, 096104 (4pp).
[40] Wendt, H.; Imarisio, G. *J. Appl. Electrochemistry* 1988, 18, 1-14.
[41] Szklarczyk, M.; Kainthio, R. C.; Bockris, J. O. *J. Electrochem. Soc.* 1989, 136, 2512-2521.
[42] Chang, R. *Mol. Cryst. Liq. Cryst.* 1974, 28, 1-8.
[43] De Gennes, P. G. *The Physics of Liquid crystals* ; Clarendon Press: Oxford, UK, 1974, chapter 3.

[44] Brodsky, A. M.; Watanabe, M.; Reinhardt, W. P. *Electrochimica Acta* 1991, 36, 1695-1697.
[45] Danielewicz-Ferchmin, I.; Ferchmin. A. R. *J. Phys. Chem.* 1996, 100, 17281-17286.
[46] Lewowski, T. *Am. J. Phys.* 1998, 66, 833-835.
[47] Semenov, V. A. *Zh. Tech. Fiz.(Russia), In Russian* 1999, 69, 127-128.
[48] Raviv, U.; Giasson, S.; Frey, J.; Klein, J. *J. Phys: Condes. Matter* 2002, 14, 9275-9283.
[49] Casagrande, Ch.; Veyssié, M. *C. R. Acad. Sc. Paris II* 1988, 306, 1423-1425.
[50] Lacrois, J.-C.; Tobazéon, R. *Appl. Phys. Lett.* 1972, 20, 251-253.
[51] Sprokel, G. J. *Mol. Cryst. Liq. Cryst.* 1973, 22, 249-260.
[52] Sussman, A. *RCA Rev.* 1974, 35, 600-612.
[53] Yamashita, M.; Amemiya, Y. *Jpn. J. Appl. Phys.* 1978, 17, 1513-1517.
[54] Sussman, A. *J. Appl. Phys.* 1978, 49, 1131-1138.
[55] Cocco, R.; Gaspard, F.; Herino, R. *J. Chim. Phys.* 1979, 76, 383-390.
[56] Bike, S. G.; Prieve, D. C. *J. Coll. Int. Sci.* 1990, 136, 95-112.
[57] Bike, S. G.; Prieve, D. C. *J. Coll. Int. Sci.* 1992, 154, 87-96.
[58] Faetti, S. *Mol. Cryst. Liq. Cryst.* 1990, 179, 217-231.
[59] Prieve, P.; Bike, S. G. *Chem. Eng. Commun.* 1987, 55, 149-164.
[60] Panzer, P.; Liu, M.; Einzel, D. *Int. J. Mod. Phys. B* 1992. 6, 3251-3278.
[61] Saranin, V. A.; Zharov, A. N.; Belonozhko, D. F. *Pis'ma Zh.T.F. (Russia), In Russian* 1997, 23, N 16, 41-44.
[62] Nasibullayev, I. Sh.; Krekhov, A. P. *Kristallografya (Russia), in Russian* 2001, 46, 540-548.
[63] Zakharov, A. V.; Dong, R. Y. *Eur. Phys. J. E* 2002, 7, 267-272.
[64] Lal, P. *Phys. Lett. A* 1986, 114, 410-412.
[65] Meyer, J.; Weihs,D. *J. Fluid Mech.* 1987, 179, 531-545.
[66] Frankel, I.; Weihs, D. *J. Fluid Mech.* 1987, 185, 361-383.
[67] Churaev, N. V. *Rev. Phys. Appl.(France)* 1988, 23, 975-987.
[68] Chen, J.-D.; Wada, N. *Phys. Rev. Lett.* 1989, 62, 3050-3053.
[69] Chen, J.-D.; Wada, N. *J. Coll. Int. Sci.* 1992, 148, 207-222.
[70] Kolb, W. B.; Cerro, R. L. *Phys. Fluids A* 1993, 5, 1549-1557.
[71] Dierking, I. *J. Phys. D.: Appl. Phys.* 2001, 34, 806-813.
[72] Yokoyama, H.; Kobayashi, Sh.; Kamei, H. *Mol. Cryst. Liq. Cryst.* 1983, 99, 39-52.
[73] Lavrentovich, O. D.; Nastishin, Yu. A. *Kristallografiya (USSR)* 1989, 34, 1529-1533.
[74] Rey, A. D. *Liq. Cryst.(UK)* 2001, 28, 549-556.
[75] Rey, A. D. *Langmuir* 2001, 17, 1922-1927.
[76] Rey, A. D. *Phys. Rev. E* 2003, 67, 011706 (14 pp).
[77] Rey, A. D. *Phys. Rev. E* 2004, 69, 041707 (9 pp).
[78] Berreman, D. In *The Physics and Chemistry of Liquid crystal Devices* ; Sprokel, G. J.; Ed.; Plenum Publ. Corp.: New York, U.S., 1980; pp 1-11.
[79] Chaure, A. *J. Mécanique (France)* 1981, 20, 333-363.
[80] Strigazzi, A. *Nuovo Cimento D* 1988, 10, 1335-1344.
[81] Thompson, P.; Robbins, M. *Physics World* 1990, N 11, 35-38.
[82] Van Alsten, J.; Granick, S. *Langmuir* 1990, 6, 876-880.
[83] Oliveira, E. A.; Figuieredo Neto, A. M.; Durand, G. *Phys. Rev. A* 1991, 44, R825-R827.
[84] Copic, M.; Clark, N. *Liq. Cryst.(UK)* 1994, 17, 149-155.

[85] Kimura, Y.; Nakano, K.; Morishita, Sh. *Proc. 6th Nordic Symposium on Tribology* 1994, 2, 313-322.

[86] Kimura, Y.; Nakano, K.; Kato, T.; Morishita, Sh. *Wear*1994, 175, 143-149.

[87] Tsoy, V. I. *Mol. Cryst. Liq. Cryst.* 1995, 264, 51-56.

[88] Tadokoro, T.; Fukazawa, T.; Toriumi, H. *Jpn. J. Appl. Phys.* 1997, 36, L1207-L1210.

[89] Mertelj, A.; Čopič, M. *Phys. Rev. Lett.* 1998, 81, 5844-5847.

[90] Kedney, P. J.; Leslie, F. M. *Liq. Cryst.(UK)* 1998, 24, 613-618.

[91] Faetti, S.; Nobili, M.; Raggi, I. *Eur. Phys. J. B* 1999, 11, 445-453.

[92] Durand, G. E.; Virga, E. G. *Phys. Rev. E* 1999, 59, 4137-4142.

[93] Stoenescu, D. N.; Dozov, I.; Martinot-Lagarde, Ph. *Mol. Cryst. Liq. Cryst.* 2000, 351, 427-434.

[94] Nasibullayev, I. Sh.; Krekhov, A. P.; Khazimullin, M. V. *Mol. Cryst. Liq. Cryst.* 2000, 351, 395-402.

[95] Jérôme, B.; Schuddeboom, P. C.; Meister, R. *Europhys. Lett.* 2002, 57, 389-395.

[96] Biscari, p. *Eur. J. Mech. B* 2002, 21, 739-750.

[97] Jánossy, I.; Kósa, T. I. *Phys. Rev. E* 2004, 70, 052701 (4 pp).

[98] Alonso, G.; Zasadzinski, J. A. *Phys. Rev. E* 2004, 69, 021602 (6 pp).

[99] Tsvetkov, V. A. *Mol. Cryst. Liq. Cryst.* 2005, 436, 203-216.

[100] Barbero, G.; Pandolfi, L. *Phys. Rev. E* 2009, 79, 051701 (8 pp).

[101] Oswald, P.; Dequidt, A.; Zywociński, A. *Phys. Rev. E* 2008, 77, 061703 (9 pp).

[102] Blanc, M.; Nespoulous, M.;Angot. E.; Nobili, M. *Phys. Rev. Lett.* 2010, 105, 127801 (4 pp).

Chapter 7

Orienting Action of an AC Electric Field with a Frequency above the Relaxational $4\pi\sigma/\varepsilon$ of the Ions on a Nematic Layer. New Methods for Measurement of the Material Parameters of Nematics and Their Application

Carr is the first author [1] who considered the orientation action of the electric field in the bulk or in isolated local regions of the nematic. When the AC electric field is with a sufficiently high amplitude it can also remove the thermal fluctuations of the nematic director [2, 3]. Delaye [4] used the attenuation of the twist fluctuations in order to measure the rotational viscosity of the nematic near the nematic-smectic-A phase transition. Dunmur et al., [5] proved that the quenching of the thermal fluctuations in the nematic phase leads to biaxiality of the liquid crystal. Galerne et al. [6], have used an additional stabilizing electric field in order to influence on the period of the chevrons. Additional stabilizing electric field is used very often in the two-frequency addressing scheme for the nematic liquid crystal displays or for the accelerating action of the AC field to shorten the relaxational process [7]. The orienting action of AC electric field was also considered by Berezin et al., [8]. The orienting electric field with a high amplitude can exert influence on a number of physical parameters of the nematics such as the viscosity [9], the alignment order S of the nematic [10] or even the value of the nematic-isotropic phase transitional temperature [11, 12]. We studied theoretically and experimentally the orienting action of the AC electric field leading to the complete orientation of the liquid crystal [13-15].

7.1. Orienting Action of AC Electric Field on an Initially-Deformed Nematic

In the practice, the liquid crystals usually contain ions and the orientation of the AC electric field is effected at a sufficiently high frequency which should be considerably above the peculiar relaxation frequency of the ions $4\pi\sigma/\varepsilon$ (in CGS units), where σ is the conductivity of the nematic and ε is the mean value of the dielectric constant. We will consider here the orienting action of the electric field, opposite to its action at the Fréedericksz transition, at which the bulk dielectric torques oppose to the surface ones. The simple methods for obtaining a planar orientation of the nematic such as the rubbing method of Zocher-Chatlain lead often to texturing of the liquid crystal with participation not only of the azimuthal angle φ but also of the polar angle θ. The initial deformation angle θ can be controlled by the application of an orienting electric field. The azimuthal angle can be controlled by the rubbing method which in addition to Zocher-Chatlain was theoretically considered by Berreman (this problem was considered in details in chapter 4) and it is convenient to consider initial sinusoidal deformations of the polar angle θ [16-21]. Until the appearance of our paper [16], there was not a theoretical model, which could permit to make qualitative conclusions about the connection between the magnitude of the initial deformations and the orienting action of the electric field. The material in this part of the work represents the experimental and theoretical study of this phenomenon for various nematic liquid crystals.

The nematics such as MBBA, EBBA, APAPA, and PAA placed between two glass plates separated by Teflon and oriented along the axis Ox by the rubbing method (see Figure 7.1) formed flat liquid crystal cells with a thickness of 10-50 μm.

An AC electric field with a frequency exceeding 200 Hz was applied along the axis Oz between the SnO_2- coated glass p;ates. At these conditions, the deviation of the nematic director from the ideal orientation is in the XOZ plane and is determined by the deformation angle θ (Figure 7.1). The liquid crystal cell was placed between crossed nicols of the polarizing microscope in such a way that the preferable axis of orientation Ox made with the crossed nicols an angle of 45^0.

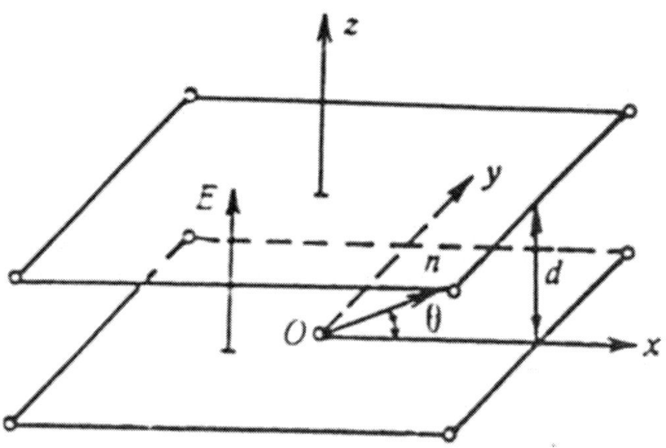

Figure 7.1. A schematic view of the liquid crystal cell used for discovering of the orienting action of AC electric field.

The light from a sodium lamp (λ = 589 μm) passing through the liquid crystal cell was recorded by a photomultiplier supplied with a potentiometer (Figure 7.2).

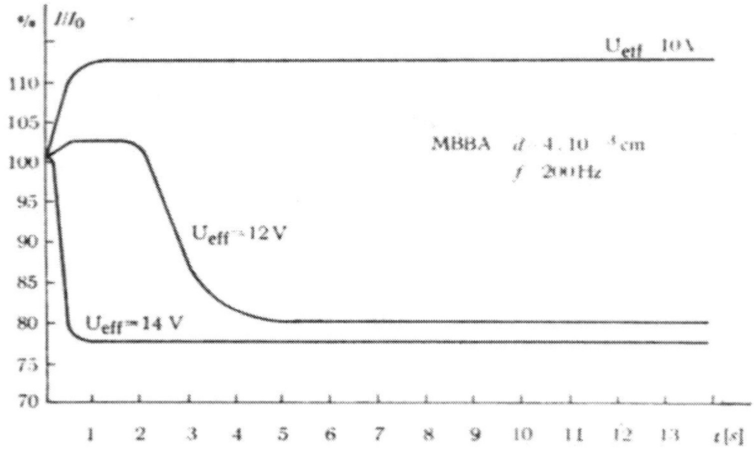

Figure 7.2. The intensity of the transmitted through the nematic light as a function of the AC voltage. The orienting action is demonstrated at a voltage of $10V_{r.m.s.}$. Domain instabilities appear at a voltageof 12V and 14V.

Applying AC voltage on a cell of MBBA with a thickness of 40 μm and initial deformation, we have observed that at a voltage of 10 $V_{r.m.s.}$, f = 200 Hz began the full orientation of the liquid crystal director. In our investigations, a voltage threshold was observed, over which a noticeable arranging action of the electric field begins as well as a value, above which the effect begins to saturate. The threshold (0.2 to 1 $V_{r.m.s.}$) and the saturation (a few volts up to 20 $V_{r.m.s.}$) voltage as well as the degree of the effect, differ and are determined chiefly by the value of the dielectric anisotropy and the magnitude of the initial deformations. The theory, developed in Reference 16, gives quite satisfactory results after comparison with the experimental results obtained by us. Let us note that when the dielectric and flexoelectric bulk torques are in opposite directions, they can be simply separated by additional application of AC electric field with sufficiently high frequency in order to stop the movement of the free ions in the samples.

7.2. Orienting Action of AC+DC Electric Fields on the Flexoelectric Domains of Vistin'-Pikin-Bobylov. Possibilities for Improvement of Their Alignment

It is well-known that the flexoelectric domains were first discovered by Vistin' [22] who, however, did not recognize their flexoelectric nature. This was done later by Bobylev and Pikin [23-26]. Further experiments confirmed the theory of Bobylev-Pikin [27, 28]. Additional experimental and theoretical studies were performed on these domains [29-32].

Finally, Ledney and Pinkevych [33-34] studied theoretically the influence of the surface anchoring on the formation of the flexoelectric domains. Marinov et al., [35, 36] considered the behavior of these flexoelectric domains under the joint action of DC and AC voltage. Marynov and Hinov showed that the additionally applied AC voltage can orient completely the flexoelectric domains as is shown in Figure 7.3 and Figure 7.4.

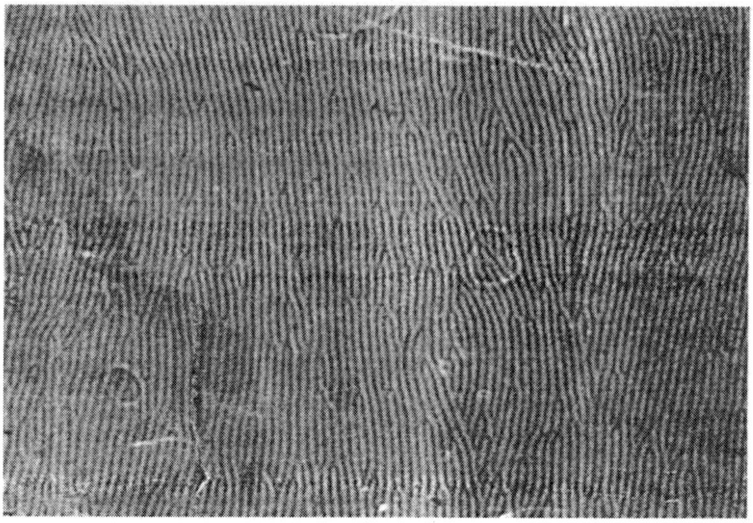

Figure 7.3. The flexoelectric domains of Vistin´-Pikin-Bobylev inder a DC voltage of 30V. It is seen many defects. Slightly uncrossed nicols, the long side of the photo corresponds to 1057 μm; room temperature.

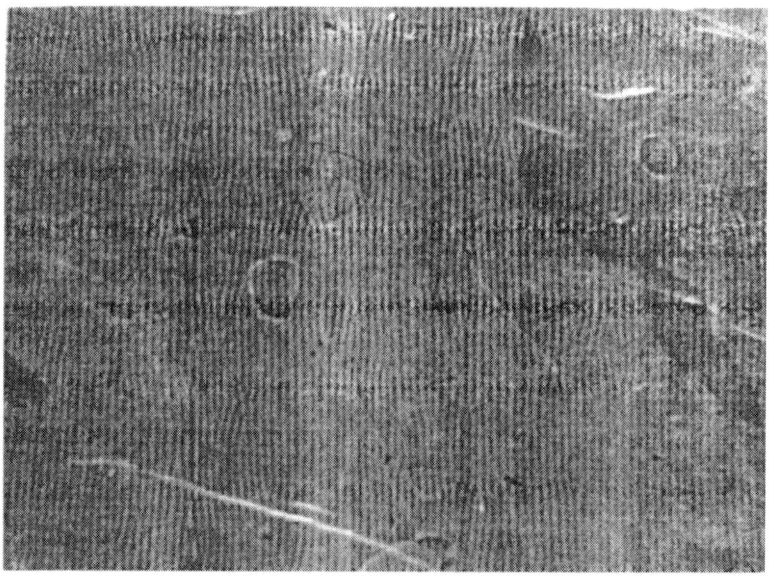

Figure 7.4. The image of Figure 7.3 with a simultaneous action of AC voltage of 25 V_{rms} (f=5kHz).

Initially we applied a DC voltage up to the appearance of the flexoelectric domains. Then we applied the AC voltage, keeping the DC voltage constant, up to the disappearance of the flexoelectric domains. Next, keeping the same value of the AC voltage, we increased the DC voltage further, just to the re-appearance of the flexoelectric domains. This procedure was performed just to obtain the pictures shown in Figure 7.3 and Figure 7.4. They indicate that the flexoelectric domains are well-aligned and can be used, for instance, for a construction of diffraction grating [37-44]. Finally, let us note that the application of two AC electric fields with different frequencies has been studied by Lev et al., [45] and new phenomena were found.

7.3. New Methods for Measurement and Application of the Flexoelectric Coefficients in Nematics

We suggest the following new methods for the determination of the sign and the magnitude of some flexocoelectric coefficients or their combinations:

a. Method for determining the sign and magnitude of ($e_{1z} + e_{3x}$) in nematics when the layer is wedge-shaped and the electric field applied is radial (this case was considered in chapter 5).
b. Method for determining the sign and the magnitude of e_{1z} in nematics: strong-weak anchored nematic layer in a transversal electric field.

Our experiment is based on that of Helfrich and co-authors [46-48] for measurement of the bend flexocoefficient e_{3x} with the difference that one uses a strong-weak anchored planar nematic layer of MBBA. The weak surface anchoring was achieved with the aid of soap. The sign of the initial tilt θ_0 can preliminarily be determined by the rubbing axis on the glass. Our tentative experimental results have shown that weak-weak anchored planar layer of MBBA is not convenient for the measurement of e_{1z} since it is very easily influenced by external influences such as temperature, vibrations, etc. Due to these reasons, we chose a strong-weak anchored MBBA layer with a surface tilt, smaller than 45^0. The set-up of the experiment is shown in Figure 7.5.

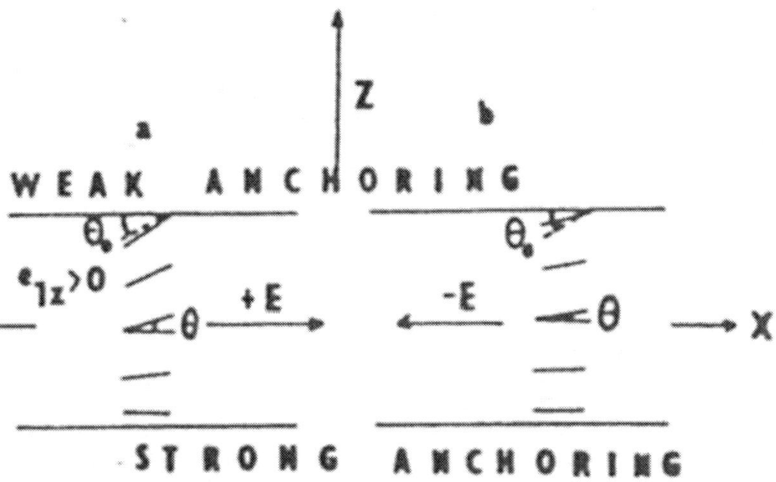

Figure 7.5. Flexoelectric splaying of a strong-weak anchored nematic layer with a slight tilt at the upper boundary: a) the flexoelectric deformations increase the initial splaying of the layer; b) the flexoelectric deformations decrease the initial splaying of the layer.

Following Helfrich and minimizing the elasto-flexo-dielectric energy of nematic, we obtain the following expression for (dθ/dz):

$$\frac{d\theta}{dz} = \frac{\left(e_{1z}E / K_{11}\right) + \left(W_{s\theta}\theta_0 / K_{11}\right)}{1 + \dfrac{W_{s\theta}}{K_{11}} - \dfrac{|\Delta\varepsilon|\ E^2 d^2}{12\pi K_{11}}} \tag{7.1}$$

where e_{1z} is the splay flexoelectric coefficient, E is the applied electric field, K_{11} is the splay elastic constant, θ_0 is the initial tilt of the nematic director, $W_{s\theta}$ is the θ-surface coupling constant and $|\Delta\varepsilon|$ is the dielectric anisotropy of the nematic.

The optical path difference between the extraordinary and the ordinary rays passing through a deformed cell is determined, in our case in the presence of an electric field, by the following expression:

$$l_2 - l_1 \cong (1/2)\, n_0 \left(\frac{n_e^2}{n_0^2} - 1\right) \left(d^3/3\right) \left(\frac{e_{1z}E}{K_{11}}\right)^2 \tag{7.2}$$

We used the following approximate formula:

$$e_{1z}^2 = \frac{l(E) - l(0)}{(1/6)\, n_e \left(\dfrac{n_e^2}{n_0^2} - 1\right) E^2 d^3} K_{11}^2 \tag{7.3}$$

The measurements were performed for four values of the applied voltage, namely, 35 V, 40 V, 50 V and 55 V. The corresponding phase difference l(E) – l(0) between the two rays was calculated according to the relation:

$$l(E) - l(0) = \frac{\Delta p \lambda}{h} = \frac{\Delta p\, 0{,}550}{2530} \tag{7.4}$$

where Δp is the difference between the positions of the micrometric screw of the microscope showing the phase difference between the two rays for each value of the electric field applied across the liquid crystal layer under study, λ=0.550 μm is the wavelength of the white light which was utilized in the experiment and h=2530 μm. In this way, we have obtained 4 values for the modulus of the splay flexoelectric coefficient e1z: $0.6\text{x}10^{-4}$ dyne$^{1/2}$, $0.85\text{x}10^{-4}$ dyne$^{1/2}$, $0.85\text{x}10^{-4}$ dyne$^{1/2}$ and $0.90\text{x}10^{-4}$ dyne$^{1/2}$. Taking into account errors, we finally obtain:

$$|e_{1z}| = (0{,}9 \pm 0{,}2)\ 10^{-4}\ dyne^{1/2} \tag{7.5}$$

The optical phase difference between the two rays, due to the flexoelectric deformations in the presence of initial tilt depended strongly on the sign of the electric field E. It is clear that for a positive sign of both the initial tilt θ_0 and the flexoelectric coefficient e_{1z}, the flexoelectric splaying leads to the increase of the initial splay when the electric field points along X and inversely, when it is directed along –X, this leads to decrease of the initial tilt (see Figure 7.5) (let us remind that the electric field is oriented from the anode (+) to the cathode (-). According to our experiment, e_{1z} should be positive:

$$e_{1z} > 0 \tag{7.6}$$

7.4. A Method for Determination of the Maximal Angle Θ_M by Means of Total Internal Reflection Measurements

It is well-known that the knowledge of the material (or physical) parameters of the liquid crystals is of enormous importance both for their scientific investigation and for their application in electronics and for displays. The method of total internal reflection of incident monochromatic light on a deformed nematic can be successfully used for the measurement of the surface angle and the surface energy as well as the maximal deformation inside the liquid crystal cells. The total internal reflection (TIR) of obliquely incident monochromatic light or a laser beam on a cell with a deformed nematic as a useful method for measurement of liquid crystal orientation has been applied for the first time in 1934 by Fréedericksz and Zwetkoff [49]. Kashnow and Stein [50] in 1973 and Labrunie and Valette [51] in 1974 were the first authors to apply the TIR method in the nematics. The technique of the TIR of a laser beam (plane monochromatic wave) obliquely incident on a cell with a nematic liquid crystal, has been used up to 1980 by several authors only: Naemura [52] in Japan and Riviére et al., [53-57] in France.

In order to solve theoretically the task for obliquely incident light beam (or a plane monochromatic light wave) on a nematic layer, it is necessary to solve the Maxwell equations with a complicated form of the dielectric tensor of the nematic, including four components, three of them – functions of the deformation angle θ. We have used in our theoretical consideration the geometric optics or wave optics approximations, which can be found in a number of books and papers [58-64]. The effort in this direction began in 1974 [65] and finished in 1986 [69]. TIR is used in many papers devoted to the display technology not only in nematics, but also in smectics (we point out several of them: [70-75]). It is necessary to mention several papers [76-83] which would be useful in the research work.

The experimental set-up is shown in Figure 7.6 where the notations are shown down below the figure.

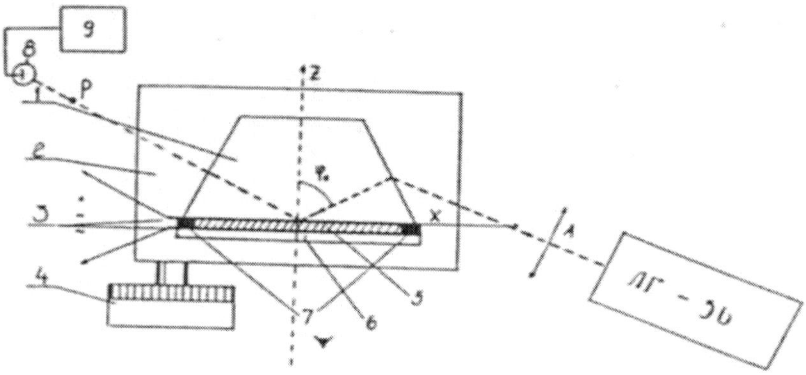

Figure 7.6. The experimental set-up for ATTR:
1- heavy glass-flint 60° prism;
2- rotational table;
3- DC or AC voltage generators;
4- a screw varying the angle of incidence ;
5- liquid crystal ;
6- usual glass plate;
7- Teflon spacers;
8- photo multiplier;
9- X-Y recorder working with a linear potentiometer.

We show schematically the TIR at geometric optics approximation and the wave optics approximation in Figure 7.7 and Figure 7.8.

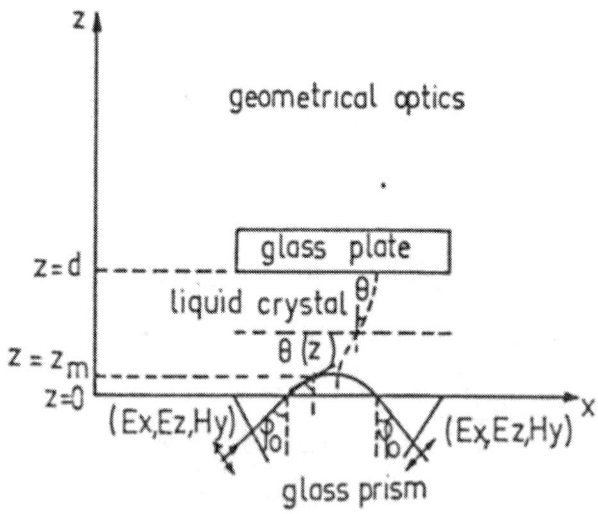

Figure 7.7. ATTR from the interface glass-flint 60° prism-deformed homeotropic nematic liquid crystal; geometrical optics.

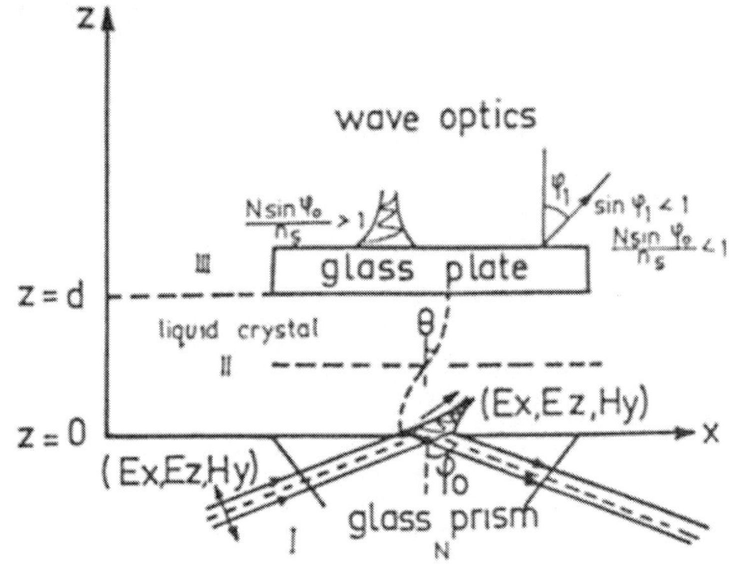

Figure 7.8. ATTR from the interface glass-flint 60° prism-deformed homeotropic nematic liquid crystal; wave optics.

We have obtained theoretically a formula including a number of parameters of the nematic under study [66]. In the approximation of the wave optics, a rough estimation shows that TIR condition is in the following form:

$$N^2 \sin^2 \varphi_0 \cong \left(n_e^2 \cos^2 \theta + n_0^2 \sin^2 \theta\right)$$
$$(\lambda_0 / b) \leq 1 \tag{7.7}$$

In this relation, N is the index of refraction of the flint-glass prism, φ_0 is the incident angle, θ is the deformation angle, n_e and n_0 are the optical indices of refraction of the liquid crystal, λ_0 is the wave length and $b = K/W_s$ is the extrapolation length introduced by de Gennes. The case when $\lambda_0/b \sim 1$ corresponds to a surface energy up to $W_s \sim 10^{-2}$ erg/cm^2. One can accept that the condition is only performed for surface energies up to $W_s \sim 5 \times 10^{-2}$ erg/cm^2.

The TIR can also be used for the determination of the maximal angle in the middle part of the liquid crystal cell in one concrete case when the surface anchoring is symmetric and when we regard a deformed planar nematic layer [68]. The comparison of the deformation at the surface and in the middle part of the cell can provide a quantitative estimate of the kind of the surface energy. Furthermore, it is possible to understand in this way the penetration of the surface forces into the nematic layer [84, 85]. In order to perform accurately such an experiment, it is necessary to take into account several requirements:

1) High quality of all the elements which enter in the experiment and first of all the heavy-glass-flint 60^0 prism (see Figure 7.6). It should be adequately polished, to be of a good material, proper flatness (without irregularities and a slope), should show a weak temperature dependence, etc.

2) High quality of the laser beam source and, consequently, of the laser beam such as collimation, estimate of its thermal energy, a size of the laser spot, etc.
3) Careful estimate of the optical properties of the liquid crystal under study, in our case MBBA.
4) It is useful to perform the experiment with observation of the laser spot, or part of the laser spot, under the polarizing microscope, This manner of performance of the experiment will increase the possibilities and the accuracy of TIR.
5) Careful estimate of the experimental errors.

The incidence angle is determined by the following formula:

$$\varphi_0 = 60^\circ \pm \arcsin\left(1/N\right)\sin\widetilde{\varphi}$$

$\widetilde{\varphi}$ is the wave incidence angle on the air-flint interface.

A substantial difference between theory and experiment is that the theory is developed for a plane wave while the experiment is performed by Gaussian laser beam. This discrepancy can be reduced with the utilization only of the central part of the laser beam, which can be achieved at lower limit of the laser power. We decided to determine the beginning of the TIR process when the laser spot transforms to a light trace at the flint-glass liquid crystal interface or when the light spot becomes much more intensive at determining of the maximal deformation inside the cell since the laser beam penetrates after that to the entire cell and then is totally reflected by it. The variations of the light spot intensities can be observed even with a naked eye. The critical angle can be determined with a high accuracy by this simple method, even with one measurement.

7.5. A Method for Determining Optical Indices of Crystals Basing on the Extended Method of Becke by Means of Nematic Liquid Crystals

We suggest here to use the Becke lines [86-91] in one modified method using nematics with appropriate optical properties (for a detail information about the Becke lines see [90]). Let us point out that the black and bright lines can interchange their position when the focus is changed as follows: black-bright – not visible – bright-black. Further, a specimen having parallel faces will produce Becke lines at the boundaries having the higher refractive index by total internal reflection. Reliable Becke lines can be observed with appropriate medium-power objectives with numerical aperture ranging from 0.5 to around 0.7. Let us note that the Becke's lines disappear when the optical indices of the crystal and the surrounding media (polar or non-polar liquid) are equal. We should point out that for the liquid crystal it is not possible since the optical indices are strongly temperature-dependent and can always be different from the optical index of the crystal. Another important requirement for observing the Becke lines is the relief which is a function of the difference between the optical indices of the crystal and those of the surrounding media. For example low relief is when this difference is smaller than 0.04, moderate relief when this difference is between 0.04 and 0.12 and high relief when Δn is bigger than 0.12. It is very easy to choose a nematic with the required optical indices. One should use a monodomain of nematic with homeotropic or planar orientation [92-97]. It is possible also to use nematics in their isotropic state.

Conclusion

1) On the basis of the simultaneous application of DC and AC an improved alignment of the flexoelectric domains of Vistin'-Pikin-Bobylov is achieved which can be used in the diffraction technique.
2) New methods for measurement of the flexoelectric coefficients of nematics are proposed.
3) For the first time it is shown that the attenuated total reflection permits the obtaining of valuable information about the maximal deformation angle iside the nematic layers.
4) Originally an improvement is suggested of the method of Becke for measurement of the optical indices of various crystals replacing the usual liquid by a liquid crystal.

References

[1] Carr, E. F. *Mol. Cryst. Liq. Cryst.* 1969, 7, 253-268.
[2] Martinand, J. L.; Durand, G. *Sol. St. Commun.* 1972, 10, 815-818.
[3] Wiener-Avnear, E. *Appl. Phys. Lett.* 1976, 29, 635-637.
[4] Delaye, M. *J. Phys. Colloq.(Paris)*1976, 37, C3-99 – C3-105.
[5] Dunmur, D. A.; Szumilin, K. *Liq. Cryst.(UK)* 1989, 6, 449-455.
[6] Galerne, Y.; Durand. G.; Veyssie, M.; Pontikis, V. *Phys. Lett. A* 1972, 38, 449-450.
[7] Wu, Sh.-T. *J. Appl. Phys.* 1985, 58, 1419-1422.
[8] Berezin, P. D.; Kompanets, I. N.; Nikitin, V. V.; Pikin, S. A. *Zh. Eksp. Teor. Fiz.(USSR)* 1973, 64, 599-607.
[9] Hess, S. *Z. Naturforsch. (a)* 1984, 39, 22-26.
[10] Palffy-Muhoray, P.; Dunmur, D. A. *Mol. Cryst. Liq. Cryst.* 1983, 97, 337-349.
[11] Savithramma, K. L.; Madhusudana, N. V. *Mol. Cryst. Liq. Cryst.* 1983, 103, 99-110.
[12] Rjumtsev, E. I.; Osipov, M. A.; Rotinyan, T. A.; Yevlampieva, N. P. *Liq. Cryst.(UK)* 1995, 18, 87-95.
[13] Sonin, A. A. *Fizika Tv. Tela (USSR)* 1984, 26, 3101-3105.
[14] Sonin, A.A.; Kaznacheev, A. B. *Fiz. Tv. Tela (USSR)* 1984, 26, 807-810.
[15] Okulska-Božek, M.; Prot, T.; Borycki, J.; Kedzierski, J. *Liq. Cryst.(UK)* 1996, 20, 349-359.
[16] Khinov, Chr. P.; Derzhanski, A. I.; Avramova, K. S. *Compt. Rend. de l'Acad. Bulg. de Sci.* 1974, 27, 1339-1342.
[17] Gu, Sh.-J.; Saha, S. K.; Wong, G. K. *Mol. Cryst. Liq. Cryst.* 1981, 69, 287-292.
[18] Aero, E. L. *Optika I Spectr. (USSR)* 1986, 60, 347-353.

[19] Barbero, G.; Beica, T.; Alexe-Ionescu, A.L.; Moldovan, R. *J. Phys. II France* 1992, 2, 2011-2024.
[20] Bryan-Brown, G. P.; Sambles, J. R.; Welford, K. R. *Liq. Cryst.(UK)* 1993, 13, 615-621.
[21] Barbero, G.; Skačej, G.; Alexe-Ionescu, A. L.; Žumer, S. *Phys. Rev. E* 1999, 60, 628-637.
[22] Vistin;, L. K. *Dokl. Ak. Nauk USSR* 1970, 194, 1318-1321.
[23] Bobylev, Y. P.; Pikin, S. A. *Zh. Eksp. Teor. Fiz.(USSR)* 1977, 72, 369=374.
[24] Bobylev, Y. P.; Chigrinov, V. G.; Pilin, S. A. *J. Phys. Colloq.(Paris)* 1979, 40, C3-331 – C3-333.
[25] Pikin, S. A. *Pis'ma Zh.T. Fiz.(USSR)* 1980, 6, 479-482.
[26] Pikin, S.A. In *Incommensurate Phases in Dielectrics*; Blinc, R.; Levanyk, A. P.; Eds.; Elsevier Science Publ. B. V., 1980; vol. 2, chap. 21, pp 320-366.
[27] Barnik, M. I.; Blinov, L. M.; Trufanov, A. N.; Umanski, B. A. *Zh. Eksp. Teor. Fiz.(USSR)* 1977, 73, 1936-1943.
[28] Barnik, M. I.; Blinov, L. M.; Trufanov, A. N.; Umanski, B. A. *J. Phys.(Paris)* 1978, 39, 417-422.
[29] Kramer, L/; Bodenschatz, E.; Pesch, W.; Thom, W.; Zimmermann, W. *Liq. Cryst.(UK)* 1989, 5, 699-715.
[30] Schiller, P.; Pelzl, G.; Demus, D. *Cryst. Res. Techn.* 1990, 25, 111-116.
[31] Cama, A.; Galatola, P.; Oldano, C.; Rajteri, M. *Mol. Cryst. Liq. Cryst.* 1995, 261, 177-185.
[32] Krekhov, A.; Pesch, W.; Buka, A. *Phys. Rev. E* 2011, 83, 051706 (13 pp).
[33] Ledney, M.; Pinkevych, I. *Zh. Eksp. Teor. Fiz.(Russia), In Russian* 2005, 127, 898-906.
[34] Ledney, M.; Pinkevych, I. *Liq. Cryst.(UK)* 2007, 34, 577-583.
[35] Marinov, Y.; Hinov, H. P.; Petrov, A. G. *J. Optoel. Ad. Mat. (Romania)* 2005, 7, 277-280.
[36] Marinov, Y. G.; Hinov, H. P. *Opt. Zh. (Russia), In Russian* 2005, 72, 70-76.
[37] Carroll, T. O. *J. Appl. Phys.* 1972, 43, 767-770.
[38] Kashnow, R. A.; Bigelow, J. E. *Appl. Optics* 1973, 12, 2302-2304.
[39] Hori, Y.; Asai, K.; Fukai, M. *IEEE Trans. El. Dev.* 1979, ED-26, 1734-1737.
[40] Chen, J.; Bos, P.; Vithana, H.; Johnson, D. L. *Appl. Phys. Lett.* 1995, 67, 2588-2590.
[41] Jepsen, M. L.; Gerritsen, H. J. *Optics Lett.* 1996, 21, 1081-1083.
[42] Zenginoglou, H. N.; Kosmopoulos, J. A. *JOSA A* 1997, 14, 669-675.
[43] Bouvier, M.; Scharf, T. *Opt. Eng.* 2000, 39, 2129-2137.
[44] Butler, J. J.; Malcuit, M. S.; Rodrigues, M. A. *JOSA B* 2002, 19, 183-189.
[45] Lev, B. I.; Sergienko, V. M.; Tomchuk, P. M.; Frolova, E. K. *Fiz. Tv. Tela. (Russia), In Russian* 1996, 38, 3490-3501.
[46] Helfrich, W. *Phys. Lett. A* 1971, 35, 393-394.
[47] Schmidt, D.; Schadt, M.; Helfrich, W. *Z. Naturforsch. (a)* 1972, 27, 277-280.
[48] Hinov, H. P.; Derzhanski, A.I. In *Liquid crystals and Ordered Fluids*; Griffin, A.; Johnson, J.; Eds.; Plenum Press: New York-London, U. S., U,K., 1984 ; vol.4, pp 1103-1121.
[49] Fréedericksz, F.; Zwetkoff, V. *Phys. Zeit. Sowjet.* 1934, 6, 490-504.
[50] Kashnow, R. A.; Stein, C. R. *Appl. Optics* 1973, 12, 2309-2311.
[51] Labrunie, G.; Valette, S. *Appl. Optics* 1974, 13, 1802-1806.
[52] Naemura, Sh. *Appl. Phys. Lett.* 1978, 33, 1-3.

[53] Rivière, D. Thèse Centre Universitare d'orsay, Paris XI. 1978.
[54] Levy, Y.; Riviére, D.; Imbert, C.; Boix, M. *Opt. Commun.* 1978, 26, 225-227.
[55] Riviére, D.; Levy. Y.; Imbert, C. *Opt. Commun.* 1978, 25, 206-210.
[56] Riviére, D.; Levy. Y.; Guyon, E, *J. Phys. Lett.(Paris)* 1979, 40, L-215 - L-218.
[57] Riviére, D.; Levy. Y. *Mol. Cryst. Liq. Cryst. Lett.* 1981, 64, 177-197.
[58] Al'pert, J.; Ginsburg, V.; Feiberg, E. *Propagation of the Radiowaves (In Russian)* Technics Press: Moscow; 1953, pp 364-383, pp 506-511.
[59] Landau, L. D.; Lifshitz, E. M. *Electrodynamics of Continuous Media*; Pergamon Press, New York, U.S., 1960.
[60] Brekhovskikh, I. M. *Waves in Layered Media*; Academic Press: New York, U.S. 1960.
[61] Hu, Ch.; Whinnery, J. R. *IEEE J. Quant. Electronics* 1974, QE-10, 556-562.
[62] Kozaki, S.; Mushiake, Y. *J. Appl. Phys.* 1975, 46, 4098-4100.
[63] Shimomura, T.; Mada, H.; Kobayashi, Sh. *Jpn. J. Appl. Phys.* 1976, 15, 1479-1484.
[64] Kozaki, S.; Harada, H, *JOSA* 1978, 68, 1592-1596.
[65] Hinov, H. P.; Sainov, S. *VI Nat. Conf. Spectroscopy*, Slanchev Bryag, Bulgaria, Sept 30-Oct. 3, 1974, Abstracts N9, 248.
[66] Hinov, H. P.; Sainov, S. *Rev. Pphys. Appl.(France)* 1980, 15, 1307-1321.
[67] Derzhanski, A. I.; Hinov, H. P.; Riviére, D. *Mol. Cryst. Liq. Cryst.* 1983, 94, 127-138.
[68] Hinov, H. P. *IV Conf. Liq. Crystals of Soc. Count.* Tibilisi 1981, E-41. 460-461.
[69] Hinov. H. P. *Phys. St. Solidi (b)* 1986, 138, 31-46.
[70] Soref, R. A. *Opt. Lett.* 1979, 4, 155-157.
[71] Soref, R. A. *Prooc. Photo Opt. Instrum. Eng.* 1979, 176, 124-132.
[72] Soref, R. A.; McMahon, D. H. *Opt. Lett.* 1980, 5, 147-149.
[73] Soref, R. A. *Opt. Lett.* 1981, 6, 275-277.
[74] Urisu, T.; Sugeta, T.; Mizushima, Y. *Appl. Optics* 1981, 20, 633-635.
[75] Soref, R. A. *Appl. Optics* 1982, 21, 1386-1392.
[76] Temple, P. A. *App. Optics* 1981. 20, 2656-2664.
[77] Chigrinov, V. G.; Korkishko, T. V.; Umanski, B. A. *Optica I Spectroscopia (USSR)* 1984, 57, 472-478.
[78] Eidner, K.; Mayer, G.; Schuster, R. *Phys. Lett. A* 1986, 118, 149-151.
[79] Eidner, K.; Mayer, G.; Schuster, R. *Phys. Lett. A* 1986, 118, 152-154.
[80] Verly, P. G. *Can. J. Phys.* 1987, 65, 476-483.
[81] Villagrań, J. C.; Thompson, J. C. *Rev. Sci. Instrum.* 1989, 60, 1201-1202.
[82] Simoni, F.; Bloisi, F.; Vicari, L.; Warenghem, W.; Ismaili, M.; Hector, D. *Europhys. Lett.* 1993, 21, 189-194.
[83] Warenghem, M.; Louvergneaux, D.; Simoni, F. *Mol. Cryst. Liq. Cryst.* 1996, 282, 235-258.
[84] Hinov, H. P. *Mol. Cryst. Liq. Cryst.* 1981, 74, 39-53.
[85] Tsvetkov, V. *Ph.D. Thesis, Institute of Radioelectronics (In Russian)*, Moscow, Russia, 1974.
[86] Donald, A. M.; Viney, C.; Windle, A. H. *Phil. Mag. B* 1985, 52, 925-941.
[87] Hartshorne, N. H.; Stuart, A. *Crystals and the Polarizing Microscope*; Elsevier: New York, 1969; pp 259-261.
[88] Hinov, H. P.; Vistin', L. K.; Marinov, Y. G. *J. Phys. Chem.* 2014, 118, 4220-4227.
[89] Williams Dorlet, C. *Défauts de Structure dans les Smectiques A*; Thèse Universite de Paris-Sud, Centre d'orsay, 1976; p 30.

[90] Davidson, M. D. *Formation of Becke Lines* In Molecular expressions: Science, Optics and You – Pionners in Optics – Friedrich Johann Karl Becke, Florida State University//micro.magnet.fsu.edu /optics/ timeline/people/becckehtml, 2013.

[91] Friedrich Johann Karl Becke (1855-1031) was best known on account of his research in the field of Rock-forming materials and how they may be determined by means of their light-reflective properties (1982, a lecture presented by Becke in Vienna University). The results of these studies were published in 1893 by the Vienna Academy.

[92] Ibrahim, I. H.; Haase, W. *J. Phys.(Paris)* 1979, 40, 191-198.

[93] Blinov, L. M.; Kovshev, E. I.; Titov, V. V. *Mol. Cryst. Liq. Cryst.* 1981, 70, 267-278.

[94] Margerum, J. D.; Lackner, A. M.; Jensen, J. E.; Miller, L. J.; Smith, W. H. Jr.; Wong, S.-M.; Van Ast, C. I. *Mol. Cryst. Liq. Cryst.* 1984, 111, 103-133.

[95] Gleeson, H. F.; Coles, H. J. *Mol. Cryst. Liq. Cryst.* 1989, 170, 9-34.

[96] Belyaev, V. V.; Chigrinov, V. G. *Appl. Optics* 1993, 32, 141-146.

[97] Dunmur, D. A. In *Physical Properties of Liquid crystals: Nematics*; Dunmur, D.; Fukuda, A.; Luckhurst. G.; Eds.; emis Data Reviews Series N 25, INSPEC, publ. IEE London, 2001; pp 315-333.

CONCLUSION

New methods and phenomena, including thermo-effects, are suggested for a deeper understanding of some interactions between the bulk and the surface of the smectics. More precisely, the smectic-A or smectic-C phases grow from a nematic or unwinded large-pitch cholesteric oriented in the middle part of the cells and from strongly-deformed nematic or unwinded large-pitch cholesteric layer in the boundary regions of the liquid crystal cells. Such a way of growing of the smectics has two advantages: first, it shows the texture fitting between the surface and bulk alignment of the smectics, sometimes unknown, and permits in some cases the approximate determination of the surface energy of smectics and second, it is useful for application in the laser-addressed liquid crystal devices. Furthermore, this way of growing of the smectic-A phase near the nematic-smectic-A phase transition can qualitatively determine the order of this, most interesting generally in physics, phase transition such as a first order, a second order, a weak first order, etc. On the other hand, the experimental results obtained with the smectic-C permit the evaluation of certain material parameters which are supplemented by some scientific results describing the behavior of the smectic-C defects in the presence of externally applied electric or magnetic fields. Finally, some of the recent investigations of the smectic-A textures show that they can be successfully applied in the photonic crystals, a modern trend in contemporary physics.

The present work includes also the electro-optic behavior of the large-pitch cholesteric at a new, strong-weak or weak-weak anchoring as well as the hysteretic behavior leading to the appearance or disappearance of various large-pitch textures at a linear scanning of a high-frequency voltage, a new electro-optic effect, important both from application and research point of view. One of the most interesting electro-optic effects, discovered in the large-pitch cholesterics, is the observation of polar unidirectional spirals under a DC voltage application. Finally, we draw attention to the discovery of a new chiral liquid crystal with unusual electro-optic behavior composed from a mixture of a nematic and a ferroelectric smectic-C* liquid crystals. This new cholesteric shows electro-optic curves with excellent rise and decay times and a high contrast of the cell. We have experimentally studied the electro-optic dynamic hysteretic behavior of weakly-anchored twisted nematic layer as well.

Flexoelectricity is also widely presented in this work. The gradient flexoelectric effect in nematics is regarded with emphasis on the two basic conditions for its observation at a very low voltage in the range of millivolts and consider the basic requirements for inhomogeneity of the electric field, the initial deformations, the weak anchoring of the liquid crystal and the role of electro-chemical material such as quinhydron which decreases the oxidation-reduction process at the two Electrodes. The flexoelectric deformations were identified except by additional application of a high-frequency orienting electric field and, qualitatively, by the simultaneous application of DC and low-frequency (several hertz) electric fields. In this way, for the first time, unusual amplification was observed of the low-frequency flexoelectric oscillations. This new effect was applied for the amplification of the flexoelectric oscillations of a homeotropic nematic layer under a transversal electric field. The „gradient flexoeffect" can be realized not only by the accumulation of ions in the nematic layer, but also by inhomogeneity of the electric field, created by the geometry of the liquid crystal cell. For example, we regarded the flexoeffect due to the wedge-shape of the liquid crystal layer. Finally, we focuss our attention on the polar flexoeffect, without taking into account the second-order elasticity.

The pulsations of the light transmitted through a nematic layer with a homeotropic anchoring and a free-slipping boundary are again considered, and new ideas are developed. The material parameters are of a great importance for the theory and application of the liquid crystals. Due to this reason, we have suggested new methods for their measurement, some cases being illustrated by examples. For instance, we have proposed a novel method for measurement of the flexoelectric coefficient of splay with a new strong-weak anchopred nematic layer. We suggested measurement of the maximal deformation angle inside the liquid crystal layer by total internal reflection (TIR) of the light passing through the liquid crystal cells. We suggested and realized a simple method with a simultaneous application of DC and high-frequency AC electric fields in order to improve the initial alignment of the flexoelectric domains of Vistin'-Pikin-Bobylev. Finally, we proposed the Becke's method for measurement of the optical index of various solids to be combined with the use of a monodomain liquid crystal in its nematic or isotropic phase.

In summary, we can qualify the work proposed as experimental, experimentally-theoretical, theoretically-experimental or theoretical depending on the topics under study.

INDEX

#

A

B

C

D

E

F

G

H

I

L

M

N

O

P

Q

R

S

T

U

V

W